SCULPTOR AND DESTROYER

SCULPTOR AND DESTROYER

Tales of Glutamate–the Brain's Most Important Neurotransmitter

MARK P. MATTSON

The MIT Press
Cambridge, Massachusetts
London, England

The MIT Press would like to thank the anonymous peer reviewers who provided comments on drafts of this book. The generous work of academic experts is essential for establishing the authority and quality of our publications. We acknowledge with gratitude the contributions of these otherwise uncredited readers.

This book was set in Adobe Garamond Pro by New Best-set Typesetters Ltd. Printed and bound in the United States of America.

Library of Congress Cataloging-in-Publication Data is available.

ISBN: 978-0-262-04818-7

10 9 8 7 6 5 4 3 2 1

To Ben Kater, who enabled and encouraged me to blaze new trails in the field of neuroscience, and to Bill Markesbery, who supported my early independent research career and from whom I learned much about Alzheimer's disease.

To the many graduate students, postdocs, and staff in my laboratories who have contributed to advancing knowledge of the neurotransmitter glutamate in brain development, plasticity, and disease.

To my wife Joanne for her love, understanding, and moral support.

Contents

Preface

This book tells the story of a simple molecule that became a master architect and commander of the brains of all animals. This remarkable molecule is glutamate. It controls the formation of the nerve cell networks as the brain develops in the womb and the various functions of those networks throughout life. Moreover, it is at the epicenter of many different neurological disorders.

When I ask laypeople to name a neurotransmitter, the most common answers are dopamine, serotonin, and "I don't know." When I ask physicians, their most common answers are dopamine, serotonin, and gamma aminobutyric acid, or GABA. Very few mention glutamate despite the fact that more than 90 percent of the neurons in the brain use glutamate as their neurotransmitter. They are "glutamatergic" neurons. The core neuronal circuitry throughout the cerebral cortex, cerebellum, hippocampus, and most other brain regions consists entirely of excitatory glutamatergic neurons and smaller numbers of inhibitory GABAergic neurons. Neurons that deploy other neurotransmitters—for instance, dopamine, serotonin, norepinephrine, and acetylcholine—are confined to only one or a few small clusters in brain structures below the cerebral cortex. These neurotransmitters can exert their effects on brain function only by modifying the ongoing activity of glutamatergic neurons.

A typical glutamatergic neuron has a pyramid-shaped central cell body, where its nucleus and genetic material reside. Extending radially from the cell body is one long axon and several shorter dendrites. At the tip of the axon and each dendrite is a motile structure called a "growth cone." During brain development, an axon's growth cone will encounter another neuron's dendrite and may form a synapse with it. When a glutamatergic neuron is activated, glutamate is released from the presynaptic terminal at the end of the axon. The glutamate then binds to specific glutamate "receptor" proteins on the dendrite of the postsynaptic neuron. This is how electrochemically encoded information flows through neuronal networks throughout the brain.

Research has shown that glutamate is the learning and memory neurotransmitter and is involved in all our feelings, thoughts, and behaviors. Our perception of everything in the environment—sights, sounds, smells, tastes, heat, cold, and so on—depend on glutamate. Memories of the past, plans for the future, imagination, creativity, language, insight, judgment, appropriate social behavior, empathy—everything that makes us who we are—involve intricately orchestrated changes in the activities of glutamatergic neuronal networks in the cerebral cortex. Glutamatergic neurons also control our body movements and influence other bodily organs, including the heart and intestines.

But glutamate also has a dark side. Abnormalities in neurons that deploy glutamate are thought to result in behavioral disorders, including autism, schizophrenia, post-traumatic stress disorder (PTSD), and depression. More dramatically, glutamate can excite neurons to death. Such "excitotoxicity" can occur rapidly in epileptic seizures, stroke, and traumatic brain injury. Excitotoxicity may also occur more insidiously in Alzheimer's disease, Parkinson's disease, amyotrophic lateral sclerosis (ALS), and Huntington's disease.

Findings from studies of organisms ranging from bacteria and plants to insects and rats show that glutamate is an evolutionarily ancient neurotransmitter. During the early evolution of animals, neurons developed elaborate treelike structures and sites of communication between neurons that were discrete and stable. These sites of electrochemical neurotransmission are called "synapses." The first neuronal circuits were for simple reflex responses and involved only two neurons: a sensory neuron that responds to a mechanical force or temperature and a motor neuron that causes a muscle to contract. As evolution proceeded, nervous systems became increasingly complex, with more neurons and more synapses, culminating in the human brain, which has about 100 billion neurons and 100 trillion synapses. Most of these synapses deploy glutamate as their neurotransmitter.

While working as a postdoc in the laboratory of Stanley (Ben) Kater at Colorado State University in 1987, I discovered the importance of glutamate in the formation of neuronal networks during brain development. By studying the growth and connectivity of newborn neurons from the brains of rat embryos, I found that glutamate controls the growth of their dendrites without affecting the axon. Glutamate released from the growth cone at the tip of a growing axon can act on a dendrite of another neuron in a manner that causes the formation of a synapse. This finding and others described in this book suggest that glutamate functions as a master "sculptor" of neuronal networks during brain development.

Throughout life, the structure of the neuronal circuits in the brain changes subtly in response to activity of the neurons in those circuits. Such changes are often referred to as "neuroplasticity." When you learn something new, the size of individual synapses and the number of synapses in circuits involved in encoding memories of that experience increase. Neuronal connections that are not used may be downsized. Glutamate controls the dynamic architecture of neuronal networks in the developing and adult brain by mechanisms involving influx of the calcium ion (Ca^{2+}) through glutamate receptor channels in the cell membrane. The Ca^{2+} then activates genes that encode proteins that promote the strengthening of the active synapse and the formation of new synapses. One such neurotrophic factor that emerges in several "tales of glutamate" told in the ensuing pages is called the brain-derived neurotrophic factor, or BDNF.

The brain consumes about 400 calories of energy during a 24-hour period, which is sufficient to keep all neurons going strong. Glutamate plays a major role in managing the production, distribution, and utilization of energy in the brain. Because of the importance of cellular energy metabolism in neuroplasticity and neurological disorders, a chapter of this book considers how glutamate controls brain "bioenergetics."

After describing the importance of glutamate in sculpting neuronal networks during brain development and in the adaptive modifications of those networks during adult life, I delve deeply into the dark "destroyer" side of glutamate. In my studies of how glutamate controls the formation of neuronal circuits during brain development, I discovered that high amounts of glutamate can kill the neurons. They can be excited to death. At the time, this process of "excitotoxicity" had just been described by John Olney and Dennis Choi (Olney 1989; Choi, Manulucci-Gedde, and Kriegstein 1987). It turns out certain naturally occurring chemicals can excite neurons to death when those chemicals are consumed. For example, in one incident in Canada, several people who had eaten shellfish at a restaurant developed an amnesia much like the short-term memory impairment of Alzheimer's disease. But most people are never exposed to such toxins, and so the question becomes whether the neurotransmitter glutamate does in fact contribute to the degeneration and death of neurons in neurological disorders.

Laboratory experiments have shown that drugs that inhibit glutamatergic synapses can protect neurons from being damaged and killed in animal models of epilepsy, stroke, and traumatic brain injury. Meanwhile, findings from studies of human patients and animal models suggest that excitotoxicity is also involved in the degeneration of neurons in many chronic neurodegenerative disorders. When neurons' ability to maintain their energy levels is compromised because of aging or genetic factors, they are particularly prone to being damaged by glutamate. Neurons are thought to experience such energy deficits and excitotoxicity in Alzheimer's disease, Parkinson's disease, Huntington's disease, and ALS.

Anxiety disorders and depression are increasingly common as people in modern societies experience chronic social stress and decreasing amounts of exercise and sleep. Brain-imaging studies suggest that there are imbalances in the excitability of certain neuronal circuits in the brain in anxiety disorders and that the same circuits are altered in depression. This may explain why treatments developed for depression are also beneficial for people with anxiety disorders. It is thought that the antidepressant and anxiolytic effects of antidepressant drugs and electroconvulsive shock are mediated by changes in the activity of glutamatergic networks and the formation of new synapses between the neurons in those networks.

Evidence suggests that autism results from accelerated fetal brain development, atypicalities in the formation of neuronal circuits, and aberrant hyperexcitability in brain regions involved in the control of social interactions. Indeed, epileptic seizures are common in children with autism, and functional magnetic resonance imaging (fMRI) studies have revealed neuronal network hyperexcitability in autistic children without seizures. In some cases, autism is caused by a gene mutation, and studies of mice engineered with such a mutation exhibit neuronal network hyperexcitability.

Psychedelic drugs such as the psilocybin in "magic mushrooms" and lysergic acid diethylamide (LSD) are thought to elicit their mind-altering effects by acting upon certain serotonin receptors on glutamatergic neurons in the prefrontal cortex. Other hallucinogenic drugs such as ketamine (special K) and phencyclidine (angel dust) directly inhibit a particular type of glutamate receptor called the "N-methyl-D-aspartate (NMDA) receptor." Drugs of abuse, including opioids, cocaine, alcohol, and nicotine, increase the amount of dopamine at synapses in the nucleus accumbens, a brain region. This increase in dopamine results from altered activity and connectivity of glutamatergic neuronal networks in other brain regions, such as the hippocampus, prefrontal cortex, and amygdala. In this way, addictive drugs cause the binging and craving experienced by their users.

The final chapter of this book considers how knowledge of glutamate's roles in neuroplasticity might be applied to the optimization of brain health. Studies have shown that people can adopt three lifestyle modifications—regular exercise, intermittent fasting, and engagement in intellectual challenges—to modulate activity in glutamatergic neuronal circuits throughout the brain in ways that enhance their performance and resilience. However, living in modern societies often results in a failure to tap the evolutionarily robust adaptive responses of neuronal networks by means of physical exertion, food deprivation, and intellectual challenges. This may explain why people with obesity are at increased risk for cognitive impairment and Alzheimer's disease.

In a very real sense, this book was written by glutamate acting within the neuronal circuits of my brain. Glutamate played a fundamental role in generating my thoughts and the words they encode, which I then transferred to a computer keyboard. It is humbling to know that compared to the eons of evolution and the construction of neuronal networks during development of my brain, the actual writing of the book seems trivial.

The purpose of *Sculptor and Destroyer* is to provide a broad perspective on glutamate's roles in brain development, neuroplasticity, bioenergetics, and neurological disorders as well as on its involvement in the brain's responses to lifestyles that maintain or do not maintain body and brain health. Readers in the fields of neuroscience, neurology, psychiatry, and psychology will dive right into the information in the pages that follow, but I have also endeavored to write the tales told in this book in a manner that may also appeal to laypeople with some knowledge of cell biology and an interest in the brain. Bon voyage!

1 INTRODUCTION

The protagonist in the drama that unfolds in the ensuing pages existed on Earth even before the emergence of the simplest cells more than 3.5 billion years ago. Also known as "glutamic acid," glutamate is one of the simplest of the 20 amino acids that cells use to construct proteins. But beyond being a building block of proteins, glutamate serves many other important functions in cells throughout the body and brain. For example, it is part of molecule called "glutathione" that squelches free radicals, and it plays an important role in the production of adenosine triphosphate (ATP)—the molecular energy currency of cells.

This book concerns a different and much more intriguing function of glutamate as the preeminent neurotransmitter in the brain. Indeed, more than 90 percent of the neurons throughout the brain deploy glutamate as their neurotransmitter. But discoveries made during the past four decades have revealed that glutamate's roles in the brain go beyond being simply a conveyor of electrochemical information between neurons. Early in brain development, even before neuronal circuits are established, glutamate is already at work controlling the growth of neurons and determining where synapses will form (figure 1.1). Once neuronal networks are formed, glutamate continues to modify their structure in ways that optimize the brain's many functions.

Numerous times during my career, colleagues have commented that they view me as a "renaissance man" because whereas most neuroscientists focus their research on one very specific problem, my work has spanned

Formation of neuronal networks during brain development

Adaptive "plasticity" of neuronal networks throughout life

Degeneration of neuronal networks in neurological disorders

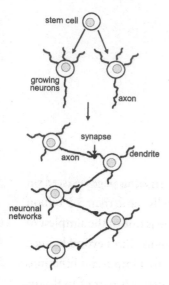
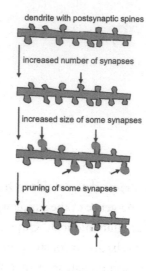
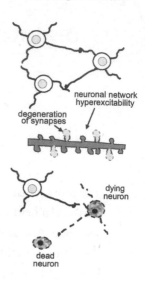

Figure 1.1

Dynamic changes in the structure of glutamatergic neuronal networks during brain development, adult life, and neurological disorders. During brain development, neurons are produced from stem cells. The neurons grow axons and dendrites and establish synapses with other neurons. Throughout life, the structure of neuronal networks changes in response to intellectual, emotional, and physical challenges. The size of individual synapses and the number of synapses may increase or decrease, and dendrites or axons may grow or regress. In many neurological disorders, glutamatergic neuronal networks become excessively active, which can result in the degeneration of synapses and in the atrophy and death of neurons.

a wide swath that includes brain evolution, development, cognition, the impact of diet and lifestyle on brain health, and the pathogenesis and treatment of neurological disorders. I am motivated to understand how the pieces of the "brain puzzle" fit together. With this inclination behind it, the purpose of this book is to provide the reader with the big picture of the expansive roles for glutamate in determining the structure and function of the brain's neuronal networks throughout life and in neurological disorders. The content is not intended to be a detailed consideration of glutamate's functions in the brain. Indeed, such a book would be orders of magnitude larger and unwieldly. The big picture would be difficult to extract from the details of the individual pieces.

OVERVIEW

The genetic code determines the specific DNA (deoxyribonucleic acid) sequence of genes and consequent production of proteins encoded by those genes. The proteins serve as the building blocks of cells and as controllers of the myriad chemical reactions by which cells multiply and function. It turns out that genes encoding the receptors for glutamate are ancient and have been discovered in simple multicellular organisms such as slime molds and mosses, and scientists have shown that glutamate controls the growth of these primitive creatures. Glutamate has also been shown to control the growth of the roots of plants and to mediate the plants' responses to stressful conditions. Therefore, very early in evolution, long before there were nervous systems, glutamate came to function as an intercellular signal that controls growth and responses to environmental stimuli. Chapter 2 describes roles for glutamate in brain evolution.

Chapters 3, 4, and 5 describe the remarkable role of glutamate as a master regulator of the structure and function of neuronal networks in the brain. The construction of a human brain involves highly intricate interactions of cells during fetal and early postnatal development. This process of brain morphogenesis begins with the proliferation of neural stem cells, from which arise billions of neurons. Each neuron extends several thin processes that grow radially away from the cell body (figure 1.2). One of the processes grows more quickly than the others and becomes much longer than the others—it is the axon. The other processes do not grow as quickly and become dendrites. The cell body of a typical neuron is about 10 micrometers (one one-hundredth of a millimeter). Dendrites may extend up to several hundred micrometers away from the cell body, and axons are much longer. Indeed, in the case of the motor neurons in the brain that control the movement of muscles, the axon is about 1 meter long.

Neurons that use glutamate as their neurotransmitter—in other words, glutamatergic neurons—are the most prevalent type of neuron in the brain.

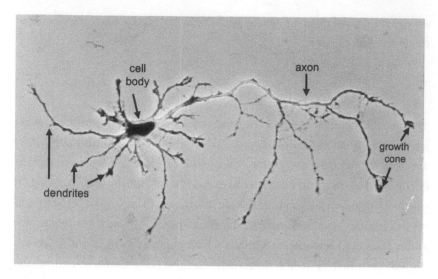

Figure 1.2

A photograph of an embryonic rat hippocampal pyramidal neuron that had been growing for three days on the surface of a culture dish. As with more than 90 percent of the neurons in the brain, this neuron is glutamatergic. When initially placed in the culture dish, the neuron was a sphere. During the ensuing three days, a single long axon with several branches and eight much shorter dendrites grew. At the tip of the axon and each dendrite is a motile structure called a "growth cone." During brain development, an axon's growth cone will encounter a dendrite of another neuron and may form a synapse with it. Adapted from figure 4.3 in Mattson (2022).

As the brain's excitatory neurotransmitter, glutamate triggers electrical impulses in the billions of neurons throughout the brain. The activity of glutamatergic neurons is kept within normal limits by the inhibitory neurotransmitter gamma aminobutyric acid (GABA). Neurons that use other neurotransmitters, such as serotonin, norepinephrine, dopamine, and acetylcholine, are few in number and are confined to small areas of the brain. Neurons that deploy these other neurotransmitters respond to glutamate and GABA and in turn fine-tune the ongoing activity in glutamatergic neuronal networks.

The complexity of the neuronal networks in the brain is truly amazing, and it is a daunting task to understand how these networks are formed and function properly (or not) throughout life. During brain development, a neuron's axon will connect to dendrites or to the cell body of other neurons

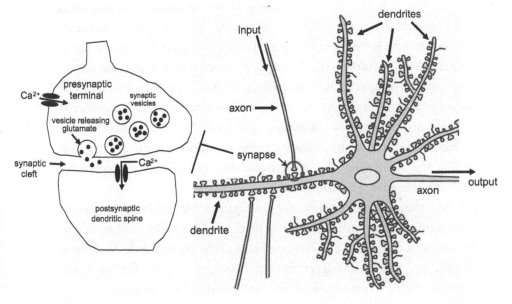

Figure 1.3

Glutamatergic synapses. A glutamatergic synapse consists of a presynaptic axon terminal that is intimately associated with a dendritic spine on a postsynaptic neuron. Glutamate is concentrated in vesicles in the presynaptic terminal. When the impulse from an action potential in the axon reaches the presynaptic terminal, Ca^{2+} enters through channels into the membrane. The Ca^{2+} influx causes the rapid fusion of synaptic vesicles with the membrane, resulting in the release of glutamate into the synaptic cleft. Glutamate binds to and thereby activates receptors on the membrane of the dendrite, resulting in membrane depolarization and Ca^{2+} influx. Each neuron throughout the brain has hundreds or even thousands of glutamatergic synapses. The image of the neuron on the right is modified from figure 1 in Smrt and Zhao (2010).

in specialized structures called "synapses." It has been estimated that there are upward of 100 trillion synapses in the human brain. An individual neuron can establish synapses with more than 1,000 other neurons.

Thousands of neuroscientists have devoted their careers to advancing an understanding of how a synapse works. Their discoveries have revealed an elegant molecular machinery that rapidly transforms an electrical impulse into a transient chemical signal at a discrete site between two neurons—a synapse. A synapse consists of the ending of a neuron's axon sending the signal and a small region of a neuron's dendrite receiving the signal (figure 1.3). The axon ending is called the "presynaptic terminal," and the part on the receiving side of a synapse is called a "dendritic spine." The very small space between the presynaptic terminal and the dendritic spine of the postsynaptic neuron is the "synaptic cleft." The size of a typical synapse is about 1 micrometer in width. When a synapse is viewed under an electron microscope, several unique features are evident. Within the presynaptic terminal are collections of small spheres, or "vesicles," located close to the membrane adjacent to the space between the axon and dendrite. These vesicles contain a neurotransmitter. In the postsynaptic membrane is an electron-dense collection of proteins called the "postsynaptic density," where the receptors for the neurotransmitter are concentrated.

At a glutamatergic synapse, the glutamate is released from the presynaptic terminal and binds to a receptor on the dendrite. The binding of glutamate to its receptor causes the membrane of the postsynaptic dendrite to "depolarize," which can trigger the postsynaptic neuron to "fire" an action potential and so to propagate an electrical impulse along its axon.

Like a battery with a positively charged end (cathode) and a negatively charged end (anode), there is a charge difference across a neuron's membrane. In the neuron's resting state, there are more positive charges on the outside of it compared to the inside of it. This is because the concentration of the positively charged sodium ion (Na^+) and of the negatively charged chloride ion (Cl^-) are greater outside the neuron than inside it. The voltage difference across the membrane is typically about −70 millivolts. Glutamate triggers depolarization and action potentials by causing the movement of Na^+ into the neuron through channels in the membrane. After the neuron fires an action potential, the voltage difference across the membrane is rapidly restored to its resting state. This "repolarization" of the membrane results from the movement of potassium ions (K^+) from the inside to the outside of the neuron through membrane channels and from the ATP (energy)-dependent "pumping" of Na^+ out of the neuron.

Throughout the brain, there are relatively small neurons that take up glutamate and convert it into the inhibitory neurotransmitter GABA, which puts the brakes on glutamatergic neurons. GABA receptors are Cl^- channels. When GABA binds to the receptor, the Cl– moves into the neuron, thereby hyperpolarizing the membrane, which reduces the neuron's excitability.

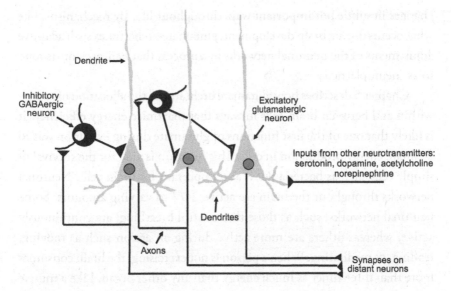

Dendrite →

Inhibitory GABAergic neuron

Excitatory glutamatergic neuron

Inputs from other neurotransmitters: serotonin, dopamine, acetylcholine norepinephrine

Dendrites

Axons

Synapses on distant neurons

Figure 1.4

The core neuronal circuitry throughout the brain consists of excitatory glutamatergic neurons and inhibitory GABAergic neurons. The axons of glutamatergic neurons synapse upon neurons within the same brain region or with neurons in other brain regions on the same or opposite side of the brain. The axons of GABAergic neurons synapse upon glutamatergic neurons within the local circuitry of the brain region in which they reside. Glutamatergic neurons also often receive inputs from serotonergic, dopaminergic, cholinergic, and noradrenergic neurons.

Presumably very early in the evolution of nervous systems, there was an adaptive advantage to being able to inhibit the activity of glutamatergic neurons. Prior to that point, the nervous systems of organisms consisted of simple reflexive responses. But for the brain to process and integrate incoming sensory information, it is important that the glutamatergic neurons be turned off after the sensory stimulus is no longer present. GABAergic neurons provide this "braking mechanism" (figure 1.4).

Acting as a neurotransmitter, glutamate is vital for the second-by-second functioning of neuronal networks. In the 1980s, however, I discovered that glutamate also "sculpts" the intricate structural features of neuronal networks during brain development and is therefore critical for the establishment of the brain's "neuroarchitecture." Other neuroscientists have shown that far from being "hard-wired," the structure of the brain's neuronal networks changes in subtle but important ways throughout life. By mechanisms like what occurs during brain development, glutamate orchestrates such adaptive adjustments of the neuronal networks in a process that neuroscientists refer to as "neuroplasticity."

Chapter 5 describes how glutamate orchestrates the allocation of energy within and between brain cells in ways that maximize energy efficiency. It is likely that one of the first functions of glutamate during evolution was to facilitate energy production in cells. This function is vital for the survival of simple cells such as bacteria but is also important for brain cells. Neuronal networks throughout the brain are active 24/7 to varying amounts. Some neuronal networks, such as those that control breathing, are continuously active, whereas others are more active during an action such as moving, reading, or meditating. When a person is not exercising, the brain consumes more than three times as much energy than any other organ. Like a muscle cell, when a neuron is very active, it uses more energy than when it is less active. Accordingly, when a neuron is stimulated by glutamate, its energy demand increases. The activation of glutamatergic synapses rapidly increases energy (ATP) production in mitochondria. Over time, activity in neuronal networks can even increase the number of mitochondria in neurons, which occurs by mechanisms similar to those that occur in muscle cells in response to regular exercise.

"It is truly amazing," wrote the French biologist François Jacob, "that a complex brain [organism], formed through an extraordinarily intricate process of morphogenesis, should be unable to perform the much simpler task of merely maintaining what already exists" (Jacob 1982). This statement reflects the theme of my four-decade career as a neuroscientist. My colleagues and I have worked to understand the cellular and molecular mechanisms that sculpt the formation of neuronal networks during brain development and why these networks are disabled or destroyed in age-related diseases such as Alzheimer's and Parkinson's. Much of this research revolved around glutamatergic neurons in cell cultures, rats, and mice.

One might intuit, as did François Jacob, that it should be easier to maintain a brain that already exists rather than design and build a new one from scratch, but, unfortunately, this is not the reality. Organisms age and die. For the individual animal, the accomplishments of millions of years of evolution are rapidly destroyed. For the species, the genetic blueprints live on and are improved upon by natural selection. Although some researchers believe that the aging process is genetically programmed, most evidence suggests that it involves an inevitable deterioration of organic molecules and the cells in which they reside. In the case of the brain, glutamate plays a role in determining whether aging occurs gracefully or results in neurodegenerative disease. Chapters 7, 8, and 9 present evidence that aberrant glutamatergic neurotransmission is a fundamental feature of a wide array of neurological disorders, including epilepsy, Alzheimer's, Parkinson's, Huntington's, stroke, anxiety, depression, and schizophrenia. In each of these brain disorders, the balance between the excitation of neurons by glutamate and their quieting by GABA is disturbed. When the balance tips heavily toward glutamate, neurons can be excited to death. This is most overt in the cases of severe epileptic seizures and stroke, but it also occurs in a more insidious manner in Alzheimer's, Parkinson's, and Huntington's diseases. The aging process itself tends to make neurons vulnerable to "excitotoxicity," and both genetic and environmental factors can influence their vulnerability, for better or worse.

Several recreational drugs exert their "mind-altering" effects by effecting changes in glutamatergic neurotransmission. Some of these drugs, such as phencyclidine (angel dust) and ketamine (special K), act directly on glutamate receptor proteins on the membrane of neurons. Others, such as the psychedelics LSD and psilocybin, exert their effects by binding to serotonin receptors on glutamatergic neurons. Addictive drugs such as opioids and nicotine also act on their own specific receptors on glutamatergic neurons. Drugs reduce anxiety by activating GABA receptors, thereby inhibiting glutamatergic neurons. Chapter 10 describes how psychoactive drugs affect glutamatergic neuronal networks in ways that result in altered behaviors.

Sculptor and Destroyer ends on a positive but cautionary note. Brain health, function, and resilience can be enhanced in several ways throughout life. These ways included regular exercise, intermittent fasting, and engagement in intellectual challenges and social interactions. Each of these lifestyle factors exerts its beneficial effects via glutamatergic neurotransmission, thereby increasing the number and size of synapses as well as the number of mitochondria in neurons. This increase occurs by a mechanism involving increased production of the neurotrophic factor BDNF. Exercise, intermittent fasting, and intellectual enrichment can also stimulate the production of new neurons from stem cells in the hippocampus in a process called "neurogenesis." In contrast, a sedentary and intellectually mundane lifestyle that includes overeating can render neuronal networks vulnerable to dysfunction and accelerated degeneration, which occurs as a result of impaired brain energy metabolism and an excitatory imbalance. Thus, glutamate can be your friend or your foe depending on your daily habits.

HISTORICAL PERSPECTIVE

By the 1940s, evidence had emerged pointing to acetylcholine as the neurotransmitter released from motor neurons and causing the contraction of muscles in vertebrate animals. But most people ignored the possibility that glutamate might be a neurotransmitter because it was known that glutamate is an amino acid and thus a building block of proteins. It was also known that glutamate is involved in several metabolic pathways and plays a critical role in the Krebs cycle by which ATP is produced in the mitochondria of cells. How could something so abundant as an amino acid involved in building proteins and energy metabolism also be a neurotransmitter?

While World War II was raging in Europe and the Pacific, Professor Takashi Hayashi was performing experiments at Keio University in Tokyo that provided the first evidence that glutamate can excite neurons (Takagaki 1996). Hayashi injected glutamate into the brain of a dog and observed that the dog exhibited epileptic seizures. Considering the social environment of imperialist Japan, Hayashi must have had considerable motivation and persistence to pursue such experiments. He was also a poet, and perhaps both his poetry writing and experiments on dogs enabled him to retreat from the turmoil of war.

At the time Hayashi published the results of his experiments, Jeffrey Watkins was a boy in high school in Australia. Watkins had a keen interest in chemistry, and after going to college in Australia, he obtained a PhD from Cambridge University in England. Watkins and the physiologist David

Curtis showed that glutamate can cause the depolarization and firing of neurons. They developed methods for recording the electrical activity of large motor neurons and smaller interneurons in the spinal cords of lightly anesthetized cats. They used recording electrodes that were placed either next to an interneuron or within a motor neuron. They found that glutamate excited both types of neurons (Curtis, Phillis, and Watkins 1960). In other experiments, they used spinal cords that had been removed from a species of toad native to Australia. The spinal cords were cut into segments to increase the access of glutamate to the neurons. A recording electrode was placed in a "ventral root," which is a large bundle of motor neuron axons. Another electrode—the "ground" electrode—was placed in the salt solution bathing the spinal cord. They found that glutamate enhanced the firing of the motor neurons (Curtis, Phillis, and Watkins 1960). According to Watkins, he had decided to see if glutamate had an effect on the neurons simply because there happened to be a bottle of glutamate in the lab.

Although Watkins did find that glutamate can depolarize neurons, he remained skeptical that it was a neurotransmitter and at the time concluded that it has a "non-specific" excitatory action upon different types of neurons.

> A transmitter function of L-glutamate did indeed seem quite unlikely in those early years. The hypothetical "receptor" would have to respond to many amino acids (either L- or D-, "natural" or "unnatural") with some general resemblance to glutamate. Furthermore, a range of glutamate enzyme inhibitors failed to affect the duration of action. Also, high concentrations were generally needed, 1000 times more than expected, say, for acetylcholine or noradrenaline at their peripheral neuroeffector sites. By the same token, we expected only very low concentrations of transmitters to be actually present in central nervous tissue. Instead, L-glutamate was among the most abundant of all small molecule constituents of brain. But the most serious objection to the transmitter possibility was that the reversal potential for L-glutamate-induced depolarisation of motoneurones was apparently different from that of the excitatory synaptic response. While the discrepancy could be explained in various ways, this result was clearly a set-back to our hypothesis, and many years were to elapse before a transmitter function for glutamate could be established. (Watkins and Jane 2006, S102)

Little did he know that future research would establish that glutamate has highly specific roles acting at synapses throughout the brain and spinal cord—actions that control all our behaviors.

Several criteria must be met for a chemical to be established as a neurotransmitter. First, it must be released from the presynaptic terminal of a neuron at a synapse. Second, it must affect the excitability of the postsynaptic neuron. Third, there must be a mechanism to remove the neurotransmitter chemical from the synapse. Fourth, it must be shown that selective blockage of the chemical's action has an effect on neuronal network activity, resulting in an alteration in one or more behaviors.

The Mexican physician and neuroscientist Ricardo Miledi made major contributions to understanding fundamental features of electrochemical neurotransmission. In 1955, he participated in a summer research program at the Marine Biological Laboratory at Woods Hole, Massachusetts, where he learned about the function of a giant axon that is part of a squid's jet propulsion system. The axon is about 2 millimeters in diameter, which provided the opportunity to record its activity using the relatively crude electrodes available at the time. After his experience at Woods Hole, Miledi took a position with Bernard Katz at the University College London. Katz had shown that acetylcholine was the neurotransmitter released from the axon terminals of motor neurons onto the frog leg muscle and that acetylcholine was released in discrete quanta. For his discoveries, Katz was awarded the Nobel Prize in 1970. In additional experiments with Miledi, Katz provided evidence that the influx of Ca^{2+} into the presynaptic terminal was required for acetylcholine release (Katz and Miledi 1965).

Miledi's early experiments utilized the frog neuromuscular preparation and electrophysiological methods in place in Katz's laboratory. But the large size of the giant axon synapses in the squid stellate ganglion provided the opportunity to answer questions not possible with frogs or rats. Stimulating and recording electrodes could be placed within the presynaptic axon terminal and postsynaptic stellate ganglion cell, respectively, rather than in vertebrate neuromuscular synapses. When Miledi stimulated the presynaptic neuron, the postsynaptic cell responded and fired an action potential. But he noticed that in the absence of stimulation of the giant axon, there were often very small discrete depolarizations of the postsynaptic ganglion cell. These "miniature postsynaptic potentials" were later shown to result from the spontaneous release of a "quanta" of neurotransmitter from an individual vesicle within the presynaptic terminal. In light of Katz's work, Miledi first studied whether acetylcholine was the neurotransmitter in the giant squid axon, but he was unable to elicit a response when he applied acetylcholine to the postsynaptic cell.

Miledi was aware that electrophysiologists had previously reported that glutamate causes the depolarization of several types of excitable cells, including crayfish muscle cells (Robbins 1958), mammalian cerebral cortical neurons (Purpura et al. 1958), and spinal cord neurons (Curtis, Phillis, and Watkins 1960). Miledi found that glutamate caused depolarization of the squid stellate ganglion cells and provided evidence that it was released from the presynaptic terminals in a quantal manner (Miledi 1967).

In the early 1970s, Solomon Snyder and his students at Johns Hopkins University showed that glutamate is concentrated in presynaptic terminals in the cerebral cortex (Wofsey, Kuhar, and Snyder 1971). Cerebral cortex tissue from rats was ground up and then centrifuged in a tube containing the sugar sucrose that was distributed in a concentration gradient. At a certain concentration of sucrose, a band of pinched-off presynaptic terminals called "synaptosomes" appeared. The synaptosomes were removed and incubated in the presence of two different amino acids, one tagged with radioactive hydrogen and the other tagged with radioactive carbon. The synaptosomes were then washed, and the amounts of radioactivity in them was determined. Among the 17 different amino acids studied, only glutamate and aspartate accumulated in the presynaptic terminals. This result showed that the presynaptic terminals have a way of actively taking up glutamate, a process that we now know involves a specific glutamate transport protein located in the presynaptic terminal's membrane.

But to understand glutamate's roles as a neurotransmitter in the brain, it became important to identify chemicals that could selectively block the presumptive receptors for glutamate. Again, scientists studying glutamate leaned on prior research on acetylcholine. It turns out that certain animals and plants have evolved the abilities to produce chemicals that block or antagonize specific types of neurotransmitter receptors.

For thousands of years, Indigenous people in Central and South America used "poison arrows" for hunting. The poison—called "curare"—was an extract of one or more plants. In the mid-1800s, the French physiologist Claude Bernard performed experiments that showed that curare interfered with the conduction of nerve impulses from motor neurons to the skeletal muscle. Then in the early 1900s, the British physiologist Henry Dale conclusively established that acetylcholine acted as a neurotransmitter at neuromuscular synapses. It was later shown that the type of acetylcholine receptor at the neuromuscular junction—the "nicotinic" receptor—is the same receptor that is at cholinergic synapses in the brain. Some animals also produce toxic substances that can paralyze other animals. For example, the Taiwanese banded krait snake's venom contains a chemical called "alpha-bungarotoxin" that irreversibly binds to and blocks the nicotinic acetylcholine receptor. This toxin is now used as a molecular probe that enables neuroscientists to visualize cholinergic synapses in animals.

Until the late 1970s, no chemicals that block neurons' responses to glutamate were known. However, several groups of researchers were identifying naturally occurring chemicals that excite neurons in a manner similar to glutamate and were much more potent than glutamate. In fact, as described in chapter 6, two such chemicals produced by algae—kainic acid and domoic acid—can excite neurons to death. Watkins, Povl Krogsgaard-Larsen, and others worked to synthesize glutamate analogues—molecules that resemble glutamate—and to determine whether they selectively activate presumptive glutamate receptors (Krogsgaard-Larsen et al. 1980). Two such analogues—NMDA and α-amino-3-hydroxy-5-methyl-4-isoxazolepropionic acid (AMPA)—were particularly potent excitants.

While Watkins was working with Tim Biscoe in England to identify glutamate receptor antagonists (Biscoe et al. 1977), so too was Hugh McLennan in Canberra, Australia (McLennan 1974). Their work resulted in the synthesis of chemicals that were highly specific antagonists of either the NMDA receptor or the AMPA receptor. This major breakthrough led to experiments that established the fundamental roles of glutamate in brain development, learning and memory, and a wide array of neurological disorders.

Another major technical advance came from Heidelberg University, where in the 1970s Erwin Neher and Bert Sakmann developed a method to record the movement of ions across the membrane of individual cells (Neher and Sakmann 1976). They called their technique the "patch clamp." It proved to be such an important advance in the field of neuroscience that Neher and Sakmann were awarded the Nobel Prize in 1991. Neuroscientists throughout the world have used patch-clamp methods to show that neurons have many different types of ion channels in their membranes. Some ion channels open in response to membrane depolarization, whereas others, including several glutamate receptor channels, open in response to the binding of the neurotransmitter to the receptor. The patch-clamp method provided a means to study the opening and closing of these different ion channels.

Patch-clamp electrodes can record the electrical currents caused by the movement of Na^+, Ca^{2+}, K^+, and Cl^- ions across the outer membrane of neurons. Glutamate causes the movement of Na^+ and Ca^{2+} into the neuron. Glutamate receptor proteins are located in the cell membrane, where they form channels through which Na^+ or Ca^{2+} ions can pass. In the absence of glutamate, the channels are closed. When glutamate binds to some types of receptor—the AMPA or kainic acid receptors—the channel opens, and Na^+ passes through the channel into the neuron. Activation of another type of glutamate receptor—the NMDA receptor—results in Ca^{2+} moving through the channel into the neuron. The reason Na^+ and Ca^{2+} move into the cell through the glutamate receptor channels is that the concentrations of these ions outside the cell are much greater than their concentrations inside the cell. The ions are soluble in water but not in fats, and because the cell membrane is made of fats, it creates a barrier for the movement of the ions. The ions can move from the outside of the cell to the inside only through the pores of ion channel proteins that are embedded in the cell membrane. The ion concentration gradients across the membrane are the result of the activity of ion pump proteins in the membrane that move the ions from the inside to the outside of the cell. The ion pumps require energy. In neurons that are very active, up to 50 percent of the cells' ATP is used to run the pumps. This is why the brain consumes so much energy.

A next important advance came when the genes that encode the different glutamate receptor proteins were identified. In 1990, the German molecular biologist Peter Seeburg and colleagues reported the identification of four genes that encode the protein subunits of the AMPA glutamate receptor (Keinanen et al. 1990). That same year Stephen Heinemann at the Salk Institute reported the cloning of genes encoding kainic acid receptors (Boulter et al. 1990), and a year later Shigetata Nakanishi at Kyoto University reported the cloning of the NMDA receptor (Moriyoshi et al. 1991). One method they used to study the ion-conducting properties of these various glutamate receptors was to inject the ribonucleic acid (RNA) encoded by the receptor gene into a frog egg and then use patch-clamp methods to determine the effects of glutamate on the movements of ions across the egg cell membrane. You may ask: Why did they use frog eggs and not neurons? There are three reasons. First, frog eggs express very few genes and do not normally respond to glutamate. Second, the frog eggs are very large, so it is relatively easy to put an electrode into them. Third, RNA can be injected into the egg, and the protein synthesis system in the egg then produces large amounts of the protein encoded by the injected RNA. When eggs were injected with RNA that encodes kainate or AMPA receptors, they became responsive to glutamate, which causes an Na^+ influx through the channel. In contrast, the NMDA receptors flux Ca^{2+}.

Humans have at least sixteen genes that code for glutamate receptor proteins. An individual receptor is formed by several glutamate receptor subunits. Some glutamate receptor subunits assemble to form the "kainic acid receptor"; others assemble to form AMPA receptors; and still others assemble to form NMDA receptors. Sodium ions pass through channels in AMPA and kainic acid receptors, whereas Ca^{2+} passes through NMDA receptor channels (figure 1.5).

In addition to the AMPA, kainic acid, and NMDA receptors, there is yet another type of glutamate receptor that is not an ion channel: the "metabotropic glutamate receptor." It has a structure similar to the receptors for dopamine, serotonin, and noradrenaline. The metabotropic glutamate

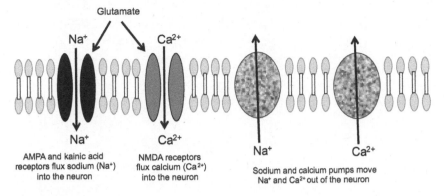

Glutamate

Na⁺ → AMPA and kainic acid receptors flux sodium (Na⁺) into the neuron

Ca²⁺ → NMDA receptors flux calcium (Ca²⁺) into the neuron

Na⁺ ↑ Ca²⁺ ↑ Sodium and calcium pumps move Na⁺ and Ca²⁺ out of the neuron

Figure 1.5

Sodium (Na⁺) and calcium (Ca²⁺) ions move into neurons through glutamate receptor channels and are moved out of neurons by "pump" proteins. There are two general types of ionotropic glutamate receptors—those that flux Na⁺ (AMPA and kainic acid receptors) and those that flux Ca²⁺ (NMDA receptors). Activation of AMPA and kainic acid receptors by glutamate causes depolarization of the membrane, which is necessary for the opening of NMDA receptors. The activities of the Na⁺ and Ca²⁺ pump proteins repolarize the membrane potential and restore intracellular Ca²⁺ levels to a basal state.

receptor protein passes through the membrane of a neuron seven times—it has seven of these "transmembrane domains" that anchor the receptor in the cell's membrane. One end of the receptor is on the outside of the membrane, and the other end is on the inside of the membrane within the neuron. When glutamate binds to the outside of the metabotropic receptor, it causes a change in the shape of a portion of the receptor inside the cell. This change enables a type of protein called the "guanosine triphosphate (GTP)–binding protein" or "G protein" to interact with the metabotropic receptor on the inner surface of the membrane. This interaction in turn activates a cascade of enzymes that can affect the function of many proteins in a cell. Some of the proteins whose functions are affected in this manner include membrane ion channels such as those for Ca²⁺ or K⁺. In this way, metabotropic glutamate receptors can fine-tune the excitability of neurons.

2 AN ANCIENT NEUROTRANSMITTER

The nitrogen in our DNA, the calcium in our teeth, the iron in our blood,
the carbon in our apple pies were made in the interiors of collapsing stars.
We are made of star stuff.
—Carl Sagan, *Cosmos*

We will probably never know where in the universe the story of glutamate
began. Possibly on Earth, but maybe not. Indeed, there are hints that glutamate may exist on other planets or their moons in our solar system. Saturn's
moon Titan is about the same size as the planet Mercury. Titan's atmosphere
is mostly nitrogen, with smaller amounts of methane and ethane. Its reddish
haze is thought to be the result of complex organic molecules falling onto the
moon's surface. Experiments have shown that when an atmosphere similar
to Titan's is exposed to ultraviolet radiation, complex molecules can form,
including nucleotides (the building blocks of DNA) and amino acids (the
building blocks of proteins). Glutamate is one such amino acid.

Glutamate probably first appeared on Earth about 4 billion years ago.
Evidence for this date came from famous experiments performed by Stanley
Miller beginning in 1952 when he was a graduate student at the University
of Chicago. Miller ran a continuous stream of steam into a mixture of oxygen, ammonia (NH_3), and methane (CH_4). He then exposed the gaseous
mixture to an electrical discharge and a week later used a method called
"paper chromatography" to detect several amino acids. Glutamate was one
of the amino acids produced in Miller's "primordial soup" experiments (D.
Ring et al. 1972).

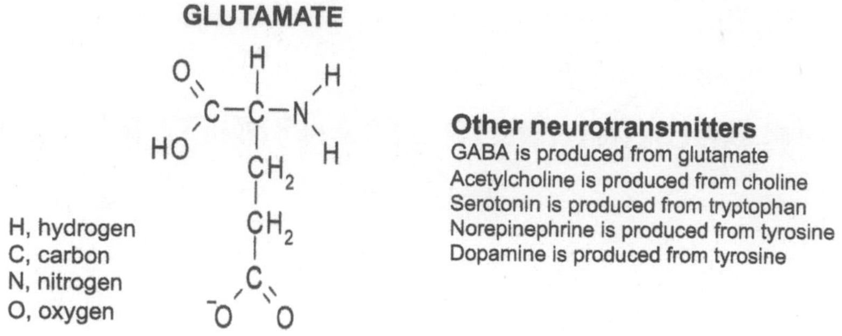

GLUTAMATE

H, hydrogen
C, carbon
N, nitrogen
O, oxygen

Other neurotransmitters
GABA is produced from glutamate
Acetylcholine is produced from choline
Serotonin is produced from tryptophan
Norepinephrine is produced from tyrosine
Dopamine is produced from tyrosine

Figure 2.1

The chemical structure of glutamate. GABA is produced from glutamate, and the other neurotransmitters are produced as indicated.

The first cells that arose on Earth about 3.5 billion years ago were bacteria. Those microbes use glutamate as a building block of their proteins. Although 20 different amino acids are now used in the production of proteins, researchers believe that very early in the evolution of life the first proteins were made from only a few amino acids. Among the 20 amino acids, glutamate has one of the simplest structures and so is thought to be one of the earliest amino acids generated in the primordial soup. Bacteria also use glutamate—in addition to its role as a component of their proteins—for the production of the ATP that keeps them alive and supports their functions and replication.

The glutamate molecule consists of five carbon atoms, four oxygen atoms, eight hydrogen atoms, and one nitrogen atom (figure 2.1). Hydrogen was the first element produced after the hot "big bang" and is the most abundant element in the universe. As the universe expanded and cooled, larger stable elements formed, and then the bonding of one element with one or more other elements to form molecules became possible. But only some of the more than 100 elements are able to form the kinds of stable bonds with other elements that are required to hold them together in molecules. The elements with such bonding capabilities include hydrogen, carbon, oxygen, nitrogen, phosphorus, and sulfur. These six elements are the building blocks of all molecules in all cells. Among all of the elements identified in the periodic table, the properties of carbon conferred it with a particularly important role in cells. The carbon atom can form four bonds at a time: it can bond with other carbon atoms or with each of the other five elements of life.

Exactly how the first cells arose on Earth is unknown, but evidence suggests they consisted of three main component molecules—RNA, proteins, and lipids (fats). The amino acid sequences of proteins were encoded by the RNA. At some point in evolution, cells synthesized DNA with base sequences complementary to those in RNA. The double-strand structure of DNA enabled sexual reproduction, which provided an advantage for more rapid evolution of species in harsh environments. In sexual reproduction, the offspring receive two copies of a gene, one from each parent. Over generations, sexual reproduction increases the diversity of genes, making it more likely that some individuals will have gene mutations that provide a survival advantage.

The sequence of amino acids in a protein is encoded by DNA. The DNA is located in the nucleus and comprises strings of nucleic acids, of which there are four types: A, adenine; C, cytosine; G, guanine; and T, thymine. Each amino acid is encoded by one or more 3-base nucleic acid sequences called "codons." Glutamate has two codons: GAA and GAG. The nucleic acid sequence of a DNA codon is used as a template for the production of messenger RNA (mRNA). Three of the four nucleic acids in RNA are the same as those of DNA (A, C, G). But instead of the T (thymine) in DNA, RNA has U (uridine). After the mRNA is produced, it is moved out of the nucleus to other locations in the cell, where amino acids are used to produce the specific protein encoded by the mRNA. Most proteins have glutamate in them. For example, insulin has 51 amino acids, of which 3 are glutamate, and the neurotrophic factor BDNF has 247 amino acids, of which 17 are glutamate. Different proteins have different functions. They may act as enzymes involved in cell metabolism; receptors for hormones, neurotransmitters, or growth factors; transcription factors; membrane transporters that move ions, glucose, and other substances into or out of cells; or antibodies produced by cells of the immune system.

At some point early in evolution, glutamate came to function as a conveyer of signals between cells. Because glutamate is soluble in water but not in fats, it cannot be released from a cell by diffusion through the lipid layer of the cell membrane. Instead, cells package glutamate in small membrane-bound spheres or "vesicles" inside of the cell. The membrane of these vesicles can then fuse with the outer cell membrane and release glutamate to the external environment. In neurons, vesicles containing glutamate are concentrated in the presynaptic terminal at the end of an axon. When that neuron is activated, the glutamate is released from the presynaptic terminal and binds to receptors on the surface of the postsynaptic neuron (figure 1.3). This binding of glutamate to its receptors is responsible for neuronal network activity throughout the brain.

Ron Petralia and his colleagues recently reviewed the evolutionary origins of synapses (Petralia et al. 2016). Structures akin to synapses are evident in a wide range of primitive multicellular organisms, including sponges and zooplankton, but it has not been determined whether glutamate is a neurotransmitter in these organisms. However, glutamatergic neurons are a major component of the nervous systems of all higher organisms, including snails, worms, insects, fish, birds, reptiles, and mammals.

GLUTAMATE PERVADES THE TREE OF LIFE

Evidence suggests that the first living creatures were bacteria, and these single-cell organisms continue to thrive in oceans, lakes, and soils throughout the planet. Some species of bacteria evolved the ability to use energy from the sun to produce sugar in a process called "photosynthesis." Then, about 2 billion years ago, plants evolved and diversified into the millions of species that now dominate the landscapes of the planet. It would be at least another billion years before animals appeared. Before we look into glutamate's roles in the brain, it is of interest to consider an evolutionary perspective on glutamate's roles in lower life forms, from bacteria and plants to slime molds, worms, and flies.

Remarkably, it was the development of patch-clamp technology and the discovery that glutamate is a neurotransmitter that led to the realization that bacteria respond to glutamate. The small size of bacteria posed a challenge in attempts to use patch-clamp methods to record ion movements across the membrane. But when scientists cleared this technical hurdle, they recorded movements of Na⁺, Ca²⁺, K⁺, and Cl⁻ across the membrane of bacteria and demonstrated the existence of channels for these ions in the membrane. Exposure of bacteria to glutamate results in the movement of K^+ ions across the membrane. This process in bacteria contrasts with the process in neurons, which instead flux Na^+ and Ca^{2+} in response to glutamate. However, there are clear similarities in the amino acid sequences of the glutamate receptor channels in bacteria and the glutamate receptor channels in neurons. These similarities in the structures of glutamate receptor ion channels suggest that the glutamate receptors in all the neurons in your brain and mine evolved from the glutamate receptors in bacteria. Although the functions of the glutamate receptor channels in the daily lives of bacteria are largely unknown, these channels may sense changes in pressure and so likely help the bacteria adapt to changes in the salt concentration of their environment. Other channels may help the bacteria tolerate acidic environments and other stressful conditions.

A type of slime mold called *Dictyostelium* lives in forests on decaying wood, where it feeds mostly on bacteria. When bacteria are abundant, *Dictyostelium* live as single cells that move around like amoeba. These amoeboid cells are haploid, which means that they have one copy of each gene in their genome. They can mate to form diploid cells, which have two copies of each gene. Diploid cells can fuse to form large cells called "plasmodia." Upon starvation, plasmodia move to the surface of the decaying wood, and approximately 100,000 individual cells come together and form a stalk, or fruiting body. Most of the cells undergo meiosis (chromosomal reduction), and the resulting haploid cells adopt a "dormant" spore state. When conditions improve and more bacteria are available, the individual spore cells become single amoeboid cells. This completes one life cycle of a slime mold.

The entire genome of *Dictyostelium* has been sequenced. Some of the genes in this primitive creature encode proteins similar to the glutamate or GABA receptors of humans (Fountain 2010). Studies have shown that slime mold cells are able to produce GABA from glutamate, just as occurs in GABAergic neurons. Interestingly, evidence suggests that GABA is produced only at certain stages of the slime mold life cycle (Y. Wu and Janetopoulos 2013). When the gene encoding a presumptive GABA receptor is deleted from the genome, the growth of amoeboid cells is increased, and spore formation is inhibited. This suggests that GABA may act as a signal in both early and late stages of the slime mold life cycle. Studies have shown that glutamate can antagonize GABA and so inhibit spore formation. Thus, very early in evolution, glutamate and GABA played important roles in "sculpting" the structure and function of multicellular organisms.

The story of glutamate and GABA in slime molds is interesting when considering the evolution of organisms with nervous systems. In the slime mold, glutamate and GABA have opposite effects on formation of the fruiting body. In nervous systems, glutamate and GABA have opposite effects on neuronal excitability. Only in some cells of slime molds and some neurons in our brains is GABA produced from glutamate. That glutamate and GABA play important roles in the formation of the multicellular fruiting body of slime molds is particularly intriguing to me. Why? Because one of the most important discoveries I made during the early years of my career was that glutamate plays an important role in the construction of neuronal networks during brain development. Chapter 3 is devoted to glutamate's "brain sculpting" function.

Mosses are a familiar site on rocks and logs in forests. These organisms are haploid and do not have seeds. Instead, after fertilization, an unbranched stalk grows, at the end of which a capsule with spores forms. Spores are released, germinate, and grow into the feltlike plants familiar to us. Sex organs form at the tips of the moss plant, and sperm swim to and fertilize eggs. Carlos Ortiz-Ramirez, Jose Feijo, and their colleagues recently provided evidence that glutamate controls both sexual reproduction and early development of the moss *Physcomitrella patens* (Ortiz-Ramirez et al. 2017). In contrast to angiosperms, which have upward of 20 glutamate receptor genes, this moss has only two genes that encode glutamate receptors. In these researchers' experiments, when these glutamate receptor genes were deleted from the sperm, the sperm swam faster but were less capable of fertilizing female gamete cells. Measurements of intracellular Ca^{2+} levels revealed lower levels in sperm lacking glutamate receptors, which is consistent with the fact that these receptors flux Ca^{2+}. Moreover, further experiments showed that after fertilization Ca^{2+} influx through glutamate receptors is necessary for the normal growth of the stalk cells from which spores form.

Moving up the tree of life, we come to plants. In some respects, plants are more complex than humans. Humans have about 25,000 genes that code for proteins, but many plants have more genes. For example, soybeans and quinoa have about 45,000 genes each, and wheat and rapeseed plants have more than 100,000 genes each. It is thought that the main reason plants have more genes is that they cannot move and thus have to be able to cope with harsh conditions, such as extreme temperatures, draught, and the chomping mouths of insects and herbivorous or omnivorous animals. To fend off insects and animals, plants produce a bewildering number of chemicals. Many genes in the plants are therefore devoted to the production of chemicals that are noxious to potential predators because they have a bitter taste or are poisonous (Koul 2005; Mattson 2015).

It was a big surprise to both plant biologists and neuroscientists when gene-sequencing studies revealed that many plants have more genes for glutamate receptors than do humans (Price, Jelesko, and Okumoto 2012)! Evidence suggests that the expression of some of the genes involved in plants' adaptive responses to stress is controlled by glutamate (Qiu et al. 2020). For example, when one leaf on a plant is damaged, other undamaged leaves increase their production of defensive chemicals that function as natural pesticides. This mechanism can protect a plant from being totally destroyed by an insect. Glutamate receptors are involved in this defensive response.

Glutamate receptors have also been shown to mediate leaf-to-leaf wound healing in plants (Mousavi et al. 2013). Masatsugu Toyota and colleagues provided evidence that when an insect damages one leaf on a plant, an increase in the Ca^{2+} level occurs in cells at the wound site (Toyota et al. 2018). Then Ca^{2+} levels increase in cells distant from the wound site in the same leaf and even in adjacent leaves on the plant. It was determined that the rate of propagation of the glutamate and Ca^{2+} signals across the plant is about 1 millimeter per second, a speed sufficient to alert distant cells that an insect is munching on another part of the plant. By manipulating genes that encode glutamate receptors, Toyota and colleagues showed that activation of glutamate receptors was responsible for propagating the Ca^{2+} signal within and between leaves. The increase of Ca^{2+} levels in leaf cells in response to glutamate causes the cells to produce chemicals that insects tend to avoid. In this way, glutamate plays an important role in plants' defensive responses to predation by insects.

A general function of glutamate in cells' and organisms' adaptive responses to stress has been conserved throughout the plant and animal kingdoms. Animals can respond much more quickly than plants to hazards they encounter. In animals, signals from sensory organs are rapidly transmitted to the brain, which then evaluates those signals in light of past experiences. Neuronal networks in the brain then send signals to muscles and other organ systems that elicit behavioral responses to the stressful situation. In animals, glutamate is the signal that transmits impulses within and between the neurons of the stress-response pathways. The entire process from perceiving a stressful situation to responding to it often takes less than one second.

The most primitive nervous systems in animals mediate reflex responses to touch. Only two types of neurons are required for such "reflex arcs"—a sensory neuron and a motor neuron. The synapse between the sensory neuron and the motor neuron deploys glutamate as its neurotransmitter. This reflex-response mechanism has been conserved throughout the evolution of animals. The classic example of such a reflex response in humans is the patellar reflex, which is often tested by physicians during a general physical examination. With the patient seated on the edge of the examination table, the doctor will use a rubber "hammer" to tap on the patellar tendon located just below the knee cap. If the reflex is normal, the leg will extend forward once and then come back to its resting position. This reflex works as follows. When the hammer strikes the tendon, it causes the tendon to stretch, which in turn stretches the endings of a sensory nerve cell. The sensory neuron responds by sending an impulse along the length of the neuron, which extends into the spinal cord. In the spinal cord, the axon terminal of the sensory neuron releases glutamate onto the dendrite of a motor neuron. The motor neuron is thus excited and sends an impulse along the length of its axon, which forms a synapse on the muscle that extends the leg. That muscle—the quadriceps femoris—contracts, and the leg moves forward.

The patellar reflex and other reflexes, such as the reflex exhibited when you touch something very hot with your hand, occur independently of the brain. There is no decision-making involved.

In insects, glutamate is the neurotransmitter deployed at the neuromuscular junction, where it stimulates muscle contraction. The glutamate receptors on the muscle cells are similar to the kainic acid receptors in neurons in the human brain. The receptors are Na^+ channels that open in response to glutamate, resulting in depolarization of the membrane and consequent influx of Ca^{2+} through voltage-dependent channels. The Ca^{2+} causes muscle contraction by stimulating the "pulling" of myosin filaments across actin filaments.

But the capabilities of the insects' nervous systems go way beyond simple reflex responses. The total number of neurons varies among insect species. For example, fruit flies have approximately 100,000 neurons, whereas honeybees have about 900,000. The brains of such insects consist of a collection of interconnected ganglia. Many of these neurons deploy glutamate as their neurotransmitter, and those glutamatergic neurons have been shown to control several behaviors. For example, studies indicate that glutamate controls circadian rhythms in fruit flies (Hamasaka et al. 2007). The capabilities of insects' brains are perhaps best illustrated by honey bees—which accurately navigate back and forth between the hive and distant food sources, and exhibit complex social interactions within the hive (Zayed and Robinson 2012).

Scientists throughout the world use the roundworm *Caenorhabditis elegans* in studies aimed at understanding the functions of genes involved in a wide array of biological processes. This worm normally lives in the soil and can easily live and propagate on agar plates in the laboratory, where their diet consists of bacteria. Several features make *C. Elegans* particularly well suited for the discovery of genes and mechanisms involved in biological processes. It is a relatively simple organism with approximately 1,000 cells. Because the worms are transparent, their development from fertilization of the egg to adulthood can be directly visualized using a microscope. Indeed, the lineages of every cell in this worm have been established, and the position of each of these cells in the adult worm is known. Individual or multiple genes can be readily disabled. Such genetic manipulations have enabled the discovery of the functions of hundreds of genes. It turns out that many *C. elegans* genes have been conserved in evolution and have homologues in humans. In several instances, discoveries of gene functions in the worms have led to the discovery of genes that have the same functions in humans. For example, the genes that control the programmed death of some cells during development were first established in *C. elegans*. Mutations of human homologues of those genes were later shown to cause cancers in humans.

The nervous system of *C. elegans* consists of 302 neurons. Most of those neurons deploy either glutamate or acetylcholine as their neurotransmitter. About 5 percent of the neurons are GABAergic, and even fewer deploy dopamine or serotonin. Working together, these neurons mediate a wide range of behaviors, including movement, the search for food, avoidance of noxious chemicals, and so on. Studies have revealed roles for glutamate in many of these behaviors, such as learning, remembering, and foraging. Takashi Kano and his colleagues identified a gene that encodes an NMDA receptor protein in *C. elegans* (Kano et al. 2008). They deleted the gene from the worms and found that the worms were then unable to learn a simple avoidance response. The worms are normally attracted to salt (NaCl). However, when the presence of salt is paired with starvation (the absence of food), the worms learn to avoid areas where salt is present. Worms lacking the NMDA receptor were unable learn to avoid the salt. The investigators then introduced the NMDA receptor gene into individual neurons of worms that lacked the gene. They found that when the gene was expressed in just two neurons, the worms were able to learn to avoid the salt. One of those neurons uses glutamate as its neurotransmitter, and the other uses acetylcholine.

Different sensory neurons in *C. elegans* respond to specific features of bacterial food, including chemicals emitted by or contained within the bacteria and the bacteria's texture. These sensory neurons are connected to interneurons in the worm's brain, and those interneurons are in turn connected to the motor neurons that control movement. When food is removed from a worm's agar plate, the worm will exhibit a local search behavior in which it explores a small area by moving forward and backward and turning. After about 15 minutes, the worm will then begin a more global search by moving forward longer distances before turning. A recent study showed that the local search behavior is initiated by two groups of glutamatergic chemosensory and mechanosensory neurons (López-Cruz et al. 2019). As a worm transitions from local to global searching, the activity of the glutamatergic sensory neurons decreases. These findings show that glutamate controls behaviors of fundamental importance for the worm's survival.

The neuronal networks that control feeding behaviors in rodents and humans are considerably more complex than those in worms. Indeed, the process of accessing food has evolved from the worms' semirandom foraging behaviors to humans' work on a job to earn money, which can then be exchanged for food at a grocery store and to purchase a refrigerator to store the food. Some readers may be surprised to learn that many brain regions that are much larger in humans than in other animals originally evolved for the purpose of maximizing success in food acquisition. As I point out my book *The Intermittent Fasting Revolution* (2022), "Prominent among the capabilities of the brain that enabled our human ancestors to overcome food scarcity are the creativity that enabled them to design and manufacture tools for hunting, to control and utilize fire, and to domesticate plants and animals; the development of languages that enabled them to accumulate vast amounts of valuable information and to pass on the information across generations; and the organization of societies via governments and religions that established distribution of effort and moral standards" (8).

Among brain regions, the prefrontal cortex has undergone the greatest increase in size during human evolution. Research by John Pearson, Karli Watson, and Michael Platt suggests that this increase in the size of glutamatergic neuronal networks in the prefrontal cortex coincided with new behaviors that enabled acquisition of high-calorie foods (Pearson, Watson, and Platt 2014). This behavioral repertoire included the invention of weapons, cooperative hunting in groups, and the use of fire to cook meat. These cognitive capabilities are mediated by neuronal networks in multiple brain regions, where the relevant information is stored, processed, and recalled and where decisions are made and appropriate actions are taken.

A BRIEF TOUR OF THE HUMAN BRAIN

It is far beyond the scope of this book to delve deeply into human brain evolution. Numerous excellent books on various aspects of this subject can easily be found on the internet. Here it is sufficient to state that the vast majority of neurons in the human cerebral cortex are glutamatergic and that it is this increase in the number of glutamatergic neurons and the glial cells associated with them that accounts for the increased size of the human brain. At this early point in the book, it is useful to have an overview of the human brain. The following "brain tour" progresses in a manner consistent with the chronology of brain evolution beginning with the primitive brainstem and culminating with the expansive frontal cortex.

The brainstem is located immediately above the spinal cord and below the rest of the brain (figure 2.2). It has three main functions. First, all information transmitted from the body to higher regions of the brain—including the cerebellum, cerebral cortex, and hippocampus—passes through the brainstem. Conversely, all information descending from higher brain regions to the body are also transmitted through the brainstem. Axons of neurons conveying sensory information from the body—pain, touch, pressure, temperature—ascend through the brainstem in the spinothalamic tract. Axons from upper motor neurons in the cerebral cortex pass through the brainstem to the spinal cord, where they form synapses on lower motor neurons. Activation of the lower motor neurons by the upper motor neurons results in the contraction of the muscles innervated by the lower motor neurons.

A second function of the brainstem involves ten cranial nerves. The cell bodies of the neurons of each cranial nerve are concentrated in "nuclei" located in discrete regions of the brainstem. The cranial nerves emanating from the brainstem include the oculomotor, trochlear, trigeminal, abducens, facial, vestibulocochlear, vagus, and hypoglossal nerves. The oculomotor, trochlear, and abducens nerves control movement of the eyes. The trigeminal nerve conveys sensations of pain, touch, and temperature of the face. It also controls contraction of muscles for movement of the jaw. The facial nerve controls muscles for facial expressions. It also receives taste information

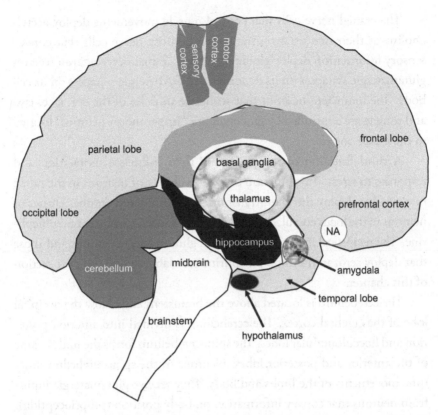

Figure 2.2
Illustration of the human brain, with many of the major regions annotated. The illustration is a medial view of the left hemisphere.

from the tongue and stimulates the salivary and tear-producing glands. The vestibulocochlear nerve innervates the ear, where it responds to the vibrations from sound. It also includes neurons that respond to the position and movements of the head as part of the vestibular system. The vagus nerve is the largest cranial nerve. The cell bodies of the vagal neurons are clustered in the lower part of the brainstem. Their axons course through the large vagus nerve to innervate the heart, intestines, and other visceral organs. Activation of vagal neurons slows heart rate and increases intestinal motility. The accessory nerve controls muscles of the neck, while the hypoglossal nerve controls movements of the tongue.

The cranial nerve cells that control muscle movements deploy acetylcholine as their neurotransmitter, whereas those nerve cells that convey sensory information deploy glutamate. Every cranial nerve neuron receives glutamatergic synapses on its dendrites and GABAergic synapses on its cell body. The brainstem neurons that stimulate muscles of the eye, face, jaw, and tongue are controlled by glutamatergic "upper motor neurons" located in the motor cortex.

A third function of the brainstem is to influence motivation and responses to stress. This function involves a cluster of neurons in the raphe nucleus that deploy the neurotransmitter serotonin and another cluster of neurons in the locus coeruleus that deploy the neurotransmitter norepinephrine. The reciprocal interactions between glutamatergic neurons and those that deploy serotonin or norepinephrine are elaborated in the next section of this chapter.

The cerebellum is located above the brainstem and below the occipital lobe of the cerebral cortex. The cerebellum is divided into anterior, posterior, and flocculonodular lobes. The spinocerebellum forms the middle zone of the anterior and posterior lobes. Neurons in the spinocerebellum fine-tune movements of the limbs and body. They receive glutamatergic inputs from neurons that convey information on body position (proprioception). They also receive inputs from visual and auditory pathways and from the trigeminal nerve. The lateral zone of the anterior and posterior lobes is the cerebrocerebellum. It receives inputs from several regions of the cerebral cortex, including the parietal lobe, and sends outputs to the thalamus and motor cortex. Neurons in the lateral zone are involved in planning movements that are about to occur by evaluating sensory information related to movements. The flocculonodular lobe plays a critical role in balance and body orientation in space.

Despite being considerably smaller than the cerebral cortex, the cerebellum contains more than half the neurons in the entire brain. One type of glutamatergic neuron, the granule cell, accounts for more than 95 percent of the neurons in the cerebellum (figure 2.3). Granule cells are much smaller than glutamatergic neurons in other brain regions. Each granule cell has only four or five short dendrites. The granule cells receive glutamatergic inputs

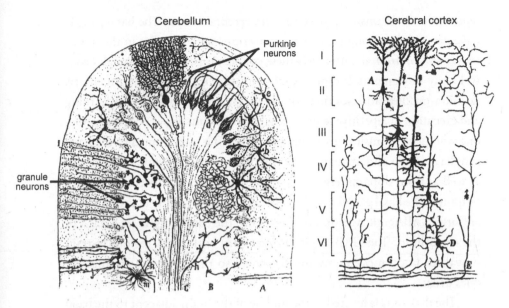

Figure 2.3
Cellular architecture of the cerebellum and cerebral cortex. The drawings were made around 1900 by the Spanish neuroscientist Santiago Ramón y Cajal. He stained small pieces of brain tissue using a method that stains only a small percentage of neurons. He then looked at them under a microscope and drew what he saw. The roman numerals denote the six different layers of the cerebral cortex.

from neurons in the cerebral cortex and spinal cord. They also receive inhibitory inputs from GABAergic interneurons located close by. The axons of the granule cells are long and have multiple short branches that form synapses on the dendrites of Purkinje cells, which are the largest neurons in the brain. Purkinje cell bodies are packed into a narrow layer, and their long, elaborate dendrites extend to the outer margin of the cerebellar lobe. Purkinje cells are inhibitory GABAergic neurons. Their axons provide the major output of the cerebellum, which functions to inhibit the descending activity of upper motor neurons not involved in a particular movement. The cerebellum also plays important roles in learning and remembering sequences of body movements such as those in playing a sport or a musical instrument.

Positioned between the brainstem and the cerebral cortex are the basal ganglia. Neurons in the basal ganglia have substantial connections with neurons in the brainstem, thalamus, and cerebral cortex. The basal ganglia have particularly strong reciprocal connections with neurons in the motor cortex and prefrontal cortex. Via these connections, the basal ganglia control body movements and play important roles in learning and memory, decision-making, and emotion. This brain region is critical for the learning and execution of repetitive sequences of body movements, as occurs in someone who plays a musical instrument or sport. The main inputs to the basal ganglia are from glutamatergic neurons in the cerebral cortex and thalamus as well as from dopaminergic neurons in the upper brainstem. The main outputs of the basal ganglia are from so-called medium spiny neurons, which are large GABAergic neurons. In this regard, the basal ganglia are similar to the cerebellum. The GABAergic outputs from both brain regions restrain unwanted body movements and behaviors so that only the appropriate ones occur.

The thalamus is located in the middle of the brain adjacent to the basal ganglia. As with the brainstem and striatum, the thalamus is an evolutionarily old structure. It functions as a relay station for sensory information. Neurons in the thalamus receive sensory inputs from glutamatergic neurons in the eyes (sight), ears (sound), tongue (taste), and spinal cord (touch, pain, pressure, temperature). Glutamatergic neurons in the thalamus then convey information associated with the sensory inputs to the appropriate sensory region of the cerebral cortex.

Research on the hippocampus (one on each side of the brain) has revealed the preeminent importance there, more than in any other brain region, of glutamate in brain development, learning and memory, and neurological disorders. The major inputs to the hippocampus are from sensory regions of the cerebral cortex—visual, auditory, somatosensory, gustatory, and olfactory. The hippocampus has strong reciprocal connections to circuits involved in emotions (amygdala), decision-making (prefrontal cortex), as well as stress responses, food-related behaviors, and sex (hypothalamus). The hippocampus is essential for learning and memory, and much of our current understanding of how memories are encoded and recalled comes from studies of glutamatergic neurons in the hippocampus. Accordingly, chapter 4 is devoted to glutamate's fundamental role in learning and memory, with a focus on the hippocampus. Neurons in the hippocampus are prone to dysfunction and damage in many brain disorders, including depression, epilepsy, and Alzheimer's disease. Chapters 6, 7, 8, and 9 tell the tales of glutamate's roles in these brain maladies.

The cerebral cortex is the most highly evolved collection of neuronal networks of the brain. Its size relative to body size has increased during evolution of primates, reaching its zenith in humans. In lower mammals, the cerebral cortex is a single "sheet" of cells that covers the underlying brain structures. During primate evolution, the cerebral cortex underwent folding, which enabled more neurons to be "packed" into the skull. This folding manifests as gyri (ridges) and sulci (grooves). Throughout the brain, the cerebral cortex is about 3 millimeters thick. At the microscopic level, neurons in the cerebral cortex exhibit a stereotypical organization (figure 2.3). Large glutamatergic pyramidal neurons are arranged in five distinct layers. As in other brain regions, GABAergic neurons are distributed throughout the cerebral cortex, where they form synapses upon adjacent glutamatergic neurons. Thus, the only neurons within the cerebral cortex are excitatory glutamatergic neurons and inhibitory GABAergic neurons.

Neuroscientists have divided the cerebral cortex into regions based on their major functions. The somatosensory cortex is a large gyrus that extends vertically in the middle of the cerebral cortex. It processes sensory information originating from sensory neurons that respond to pain, touch, pressure, and temperature and that are distributed throughout the limbs and body. Chemoreceptors that respond to chemicals in the air—odors—are located in the nose. They are innervated by neurons in the olfactory bulb, which relay the information to a region of the cerebral cortex known as the uncus. The uncus is located in the inferior part of the frontal lobe and is closely connected to brain regions critical for learning and memory (hippocampus) and fear responses (amygdala). In contrast to the other senses, the olfactory information goes directly to the olfactory cortex and does not pass through the thalamus.

The most posterior region of the cerebral cortex is the occipital lobe or visual cortex. It receives information from the eyes. The axons of the ganglion neurons of the retina extend through the optic nerve to neurons in the thalamus. The thalamic neurons relay the signals coming from the eye to neurons in the visual cortex. This information undergoes initial processing by glutamatergic neurons in the visual cortex and is then transmitted to glutamatergic neurons in other brain regions, including the parietal cortex, frontal cortex, and hippocampus.

The auditory cortex is located in the upper part of the temporal lobe. It receives inputs from the neurons that respond to the vibrations of the eardrum caused by sound waves in the air. Eardrum vibrations are transmitted to three small bones, which in turn cause movements of fluid in a spiraling tube, the cochlea. The fluid movements stimulate hair cells that line the cochlea, and the hair cells transmit the signal to neurons in the spiral ganglion. The axons of the glutamatergic spiral ganglion neurons project to the thalamus and are then relayed via glutamatergic neurons to the auditory cortex.

Not surprisingly, the brain's "language centers" include regions of the cerebral cortex adjacent to the visual cortex, auditory cortex, and motor cortex: a region of the inferior parietal cortex adjacent to the visual cortex, a region of the superior temporal lobe adjacent to the auditory cortex, and a region of the frontal lobe immediately adjacent to the motor cortex. As might be expected, the size of these language centers is much greater in humans than in any other living primate.

There is a rich history of research on language by anthropologists, behaviorists, neuroscientists, and linguists. In 1861, the French neurologist and anthropologist Paul Broca described the symptoms of a patient who had suffered damage to a region of the frontal cortex adjacent to the motor cortex. The patient was able only to correctly pronounce one word, *tan*. Another patient had damage to the same brain region and exhibited a severe problem in articulating words. This brain region, which is called "Broca's area," is now established as the speech center. Glutamatergic neurons in Broca's area receive inputs from neurons in the auditory and visual cortexes and send outputs to motor neurons that control muscles of the larynx and tongue. In 1874, the German physician Carl Wernicke was studying patients with aphasia, which is the loss of the ability to comprehend and/or express verbal or written language. Patients who suffer a stroke can exhibit aphasia.

Wernicke described a brain region in the posterior part of the upper gyrus of the temporal lobe that when damaged results in aphasia. Since then, adjacent regions of the parietal cortex have also been shown to be involved in speech comprehension and expression.

In addition to the language areas, a brain region that has increased in size dramatically during primate evolution is the prefrontal cortex. In fact, the increase in the size of the prefrontal cortex accounts for much of the difference in overall brain size between humans and our most closely related living primate, the chimpanzee. The prefrontal cortex, which is located at the very front of the frontal lobe is critical for intelligence, creativity, language processing, and decision-making. It enables humans to be adept at efficiently processing information in ways that assist them in being successful in achieving their goals. Interestingly, the prefrontal cortex does not become fully developed until a person is about 20 years old. Other brain regions are fully developed much earlier in life. Neuroscientists believe that this fact may account for the poor decision-making of most children and adolescents. The prefrontal cortex is also of interest with respect to mind-altering drugs and psychiatric disorders. Chapter 9 describes how alterations in prefrontal cortex glutamatergic neurons contribute to the symptoms of schizophrenia, and chapter 10 explains how alterations in glutamatergic neurotransmission in the prefrontal cortex are involved in the mind-altering effects of hallucinogenic drugs.

DOPAMINE, SEROTONIN, AND OTHER UNDERLINGS

From an evolutionary perspective, it is of considerable interest that all of the major neurotransmitters in the brain are either amino acids or metabolites of amino acids. Dopamine and noradrenaline are produced from the amino acid tyrosine; serotonin is produced from the amino acid tryptophan; and acetylcholine is produced from the amino acid serine. The amino acids aspartic acid and serine can also affect neuronal activity but are not generally considered neurotransmitters in that they are not concentrated in synaptic vesicles and released locally at synapses. However, aspartic acid and serine can enhance the activation of glutamate receptors.

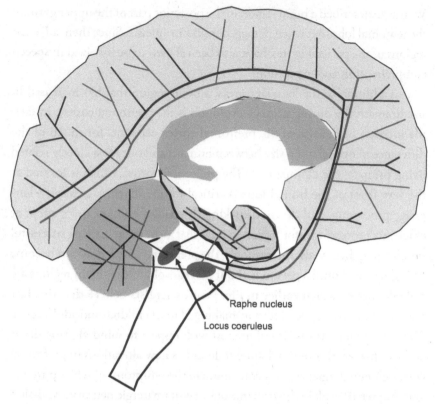

Raphe nucleus

Locus coeruleus

Figure 2.4
Serotonergic and noradrenergic neurons are located in discrete areas of the brainstem and project their axons to neurons throughout the brain. Neurons that deploy serotonin are located in the raphe nucleus (*gray*). Neurons that deploy norepinephrine are located in the locus coeruleus (*black*). The axons of these neurons form synapses with glutamatergic neurons throughout the brain.

In contrast to glutamatergic and GABAergic neurons, which are present in all regions of the brain, neurons that produce dopamine, serotonin, noradrenaline, or acetylcholine are produced by neurons only in one or a few brain regions. Dopaminergic neurons are located in two relatively small regions of the brain, the substantia nigra and the ventral tegmental area (VTA) located just above the brainstem. Neurons that produce serotonin or norepinephrine are located in the brainstem just above the spinal cord (figure 2.4). Cholinergic neurons, those that deploy acetylcholine, are located below the frontal part of the cerebral cortex in an area called the "basal forebrain." Although the cell bodies and dendrites of cholinergic, serotonergic, noradrenergic, and dopaminergic neurons are confined to discrete areas of the brain, these neurons have very long and branching axons that extend throughout the cerebral cortex. Their axons form synapses with glutamatergic neurons in all brain regions.

The only way that serotonin, norepinephrine, dopamine, acetylcholine, and GABA can affect cognition, emotions, and behaviors is by acting upon glutamatergic neurons (figure 1.4). Moreover, the activities of all neurons that use these other neurotransmitters are controlled by glutamatergic neurons. Glutamate is their master and commander. For these reasons, I refer to these other neurotransmitters as "underlings." This is not to say that the underling neurotransmitters are unimportant. Quite the contrary, they play critical roles in modulating the activity of glutamatergic neurons in ways that enable optimal brain function and adaptations of neuronal networks to environmental challenges.

The cholinergic neurons in the basal forebrain receive inputs from glutamatergic neurons in the entorhinal cortex, hippocampus, and olfactory cortex. The outputs of the cholinergic neurons project to the hippocampus and throughout the cerebral cortex. Basal forebrain cholinergic neurons play important role in wakefulness, attention, and learning and memory. There are two types of receptors for acetylcholine: nicotinic acetylcholine receptors are Na^+ channels; muscarinic cholinergic receptors are similar to metabotropic glutamate receptors in that their activation stimulates Ca^{2+} release from intracellular stores and activates certain protein kinases.

The serotonergic neurons in the brainstem raphe nucleus receive glutamatergic inputs from many brain regions, including the brainstem, prefrontal cortex, cingulate cortex, and hypothalamus. Through these connections, the serotonergic neurons are influenced by activity in circuits involved in the regulation of sleep, decision-making, and hormone production. The serotonergic neurons have very long axons. Some of these axons project upward to higher regions of the brain, and others project downward to the spinal cord. Axons projecting upward innervate glutamatergic and GABAergic neurons in essentially all regions of the brain. These serotonergic inputs can modulate the activity and plasticity of the glutamatergic neurons. For example, evidence suggests that serotonin can enhance hippocampal synaptic plasticity in ways that improve cognition; affect GABAergic and glutamatergic neurons in the amygdala in ways that reduce anxiety; and modulate the activity of neurons in the hypothalamus that control circadian rhythms and sleep. Raphe serotonergic neurons also project to neurons in the dorsal region of the spinal cord. Evidence suggests that activation of the serotonergic neurons that innervate the spinal cord can inhibit pain. This may explain why antidepressant drugs that increase synaptic serotonin levels can be effective in treating patients with chronic pain.

Fourteen different serotonin receptors have been identified. They all have a similar structure, with seven regions that pass through the membrane. Activation of the different serotonin receptors affect glutamatergic neurons in three main ways. The first is by increasing the production of a chemical called "cyclic AMP"; the second is by decreasing the production of cyclic AMP; and the third is by causing an increase in Ca^{2+} levels in the neuron. The receptors that increase cyclic AMP levels play particularly important roles in serotonin's ability to enhance learning and memory. Serotonin has been shown to play roles in the behaviors of evolutionarily ancient animals with simple nervous systems, such as snails and leeches. In such animals, serotonergic neurons control feeding behaviors and the learning of various tasks in laboratory experiments. In fact, the neuroscientist Eric Kandel received the Nobel Prize in Physiology or Medicine in 2000 for his discovery that serotonin plays a key role in learning and memory by increasing cyclic AMP levels in certain neurons in a sea slug called *Aplysia*.

Many readers may be familiar with epinephrine, which is also known as adrenaline. Epinephrine is produced by cells located in the center or "medulla" of the adrenal gland. In response to stressful situations, epinephrine is released into the blood and circulates throughout the body. It causes an increase in heart rate, blood pressure, and blood flow to the muscles. Neurons in the locus coeruleus deploy the neurotransmitter norepinephrine, which is closely related to epinephrine. Norepinephrine plays prominent roles in behavioral and neuroendocrine responses to stress. Noradrenergic neurons in the locus coeruleus receive glutamatergic inputs from the amygdala, hypothalamus, brainstem circuits called the "reticular activating system," and several regions of the cerebral cortex. Through these connections, the noradrenergic neurons are activated during stressful situations. Their activation then amplifies the stress response. Norepinephrine influences the autonomic nervous system in ways that increase activity of the sympathetic neurons that function to increase heart rate and blood pressure. Particularly important in the brain's responses to stress are noradrenergic inputs to brain regions involved in escape responses, attention, memory, and emotions—including the thalamus, cerebellum, amygdala, hippocampus, prefrontal cortex, and hypothalamus. Studies have shown that norepinephrine enhances the memories of stressful events via receptors that increase cyclic AMP levels in glutamatergic neurons in the hippocampus.

In popular culture, dopamine is known as the neurotransmitter of pleasure and addiction. More broadly, dopamine is involved in motivation, reinforcement of pleasurable experiences, and reward. Dopaminergic neuronal networks are often referred to as "reward circuits." These reward circuits evolved for the purpose of enabling animals to determine whether an action is desirable or aversive—for example, whether to eat a particular type of berry or not, whether to have sex with a particular individual or not. To better define such behaviors, neuroscientists use the term *motivational salience*, which can be defined as paying attention to and cognitively processing a stimulus so as to propel behavior toward or away from an object or perceived outcome or event. Dopamine exerts its effects on motivational salience by modulating the activity of glutamatergic neurons in the prefrontal cortex, basal ganglia, and hippocampus.

Neuroscientists and neurologists have elucidated neuronal circuits involved in motivated behaviors, reward, and addiction (figure 2.5). A very important component of these circuits is a collection of dopaminergic neurons in the VTA, which is located at the upper end of the brainstem. These dopaminergic neurons receive glutamatergic inputs from the prefrontal

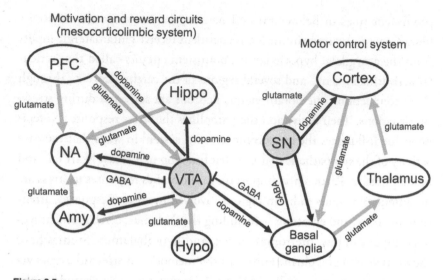

Figure 2.5

Diagrams showing interconnections of dopaminergic neurons and the brain regions that control motivation and reward as well as body movements. Amy = amygdala; Hippo = hippocampus; Hypo = hypothalamus; NA = nucleus accumbens; PFC = prefrontal cortex; SN = substantia nigra; VTA = ventral tegmental area.

cortex, hypothalamus, and thalamus. They also receive GABAergic inputs from the hypothalamus and the nucleus accumbens, a region of the basal forebrain. The dopaminergic neurons in the VTA mainly innervate GABAergic neurons in the nucleus accumbens. Glutamatergic and GABAergic neurons in the nucleus accumbens receive glutamatergic inputs from the several brain regions, including the prefrontal cortex, hippocampus, and amygdala. In addition to innervating the nucleus accumbens, the dopaminergic neurons in the VTA send outputs to the hippocampus, amygdala, and prefrontal cortex.

What are the evolutionary "forces" that sculpted the brain's motivation and reward circuits? The two most obvious answers are that these circuits evolved to enable success in the acquisition of food and sex. Eating tasty food and having sex are not only among the most pleasurable experiences in life but also the most important behaviors for survival and the passing of one's genes on to the next generation. Advancements in technology have led to increased opportunities for engaging in behaviors that can become addictive.

Together with a lack of exercise, overindulgence in highly palatable foods has become the major risk factor for the development of chronic diseases and early death from cardiovascular disease, diabetes, and cancers. Obesity is most often the consequence of an addiction to processed foods containing high amounts of sugar and saturated fats. Just as easy access to obesogenic foods increases the likelihood that a person will become addicted to them, ready access to pornographic materials on the internet increases the likelihood of addiction to them.

Beyond its important roles in motivation and reward, dopamine is involved in the second-by-second control of body movements. Whereas the dopaminergic neurons involved in motivation and reward are located in the VTA, those controlling body movements are located in an adjacent region of the midbrain called the "substantia nigra." It is estimated that the axon of each dopaminergic neuron in the substantia nigra forms more than 300,000 synapses on its target neurons in the striatum. Dopaminergic neurons in the substantia nigra exhibit a high level of spontaneous activity. This activity is normally constrained by strong inhibitory inputs from GABAergic medium spiny neurons in the basal ganglia.

Among the underling neurotransmitters, GABA is particularly important. GABAergic neurons are present throughout the cerebral cortex, hippocampus, cerebellum, and other brain regions. Their main function is to provide rapid feedback inhibition of glutamatergic neurons so as to prevent aberrant hyperactivity in glutamatergic circuits. Without the GABAergic brakes, glutamatergic neurons would "fry" themselves. This fact has been revealed by monitoring the activities of glutamatergic neurons in cell cultures. Approximately 90 percent of the neurons in cultures of embryonic rodents' cerebral cortex or hippocampus are glutamatergic, and the remainder are GABAergic. As the glutamatergic neurons establish synaptic connections, they exhibit spontaneous electrical activity, which can be recorded using microelectrodes. The activities of dozens of neurons can be simultaneously observed using molecular probes that enable imaging of intracellular Ca^{2+} levels in the neurons. Such studies have shown that individual glutamatergic neurons exhibit oscillations in their activity, that is to say, repeating sequences of activity among neurons in the network. All such activity ceases when a drug that blocks glutamate receptors is added to the medium that is bathing the neurons. Conversely, when a drug that blocks GABA receptors is added to the cultures, the glutamatergic neurons' activity increases dramatically and appears chaotic, as in what happens during an epileptic seizure. Chapters 7 and 8 focus on the roles of glutamate in the neuronal network hyperexcitability that occurs not only in epilepsy but also in Alzheimer's disease, ALS, and other neurological disorders.

3 SCULPTING BABY'S BRAIN

The bewildering complexity of the human brain's neuronal networks is established by the intricate processes of cellular morphogenesis. These processes begin with fertilization of an egg in one of mother's fallopian tubes, which connect her ovaries to her uterus. The fertilized egg divides several times as it is moved into the uterus, where cell numbers continue to increase exponentially. By the end of the first month of pregnancy, the fetus's nascent brain can be seen with the naked eye. By three months, the developing cerebral cortex is evident and has a smooth surface. During the remainder of gestation, the cerebral cortex grows in size and develops the folds that characterize the adult brain. This folding enables more neurons to be "packed" within the volume allotted by the skull. From birth to the grave, a person's brain will contain approximately 90 billion neurons, of which about 20 billion are in the cerebral cortex and 50 billion in the cerebellum. Altogether, more than 80 billion of the neurons and 90 trillion of the synapses in the brain are glutamatergic.

As the brain develops, neuronal stem cells proliferate and then stop dividing and become neurons. Stem cells are spherical in shape. Neurons born from the stem cells are initially spheres, but within a few hours they begin extending several processes called "neurites." Within 24 to 48 hours, one of the neurites begins to grow much more rapidly than the others and will become the axon (figure 1.2). The other neurites grow more slowly and branch considerably to form a "dendritic tree." The axon continues to grow until it encounters other neurons that it has the opportunity to form synapses upon. These "target neurons" for the axon may be within the same brain region or a different brain region or the spinal cord. Success in finding a target neuron and forming a synapse on the target neuron is essential for a neuron to survive during brain development. Neurons that are not successful in forming synapses will die, and many neurons do die during brain development. It is thought that this initial overproduction of neurons ensures that the brain contains a sufficient number of neurons to form all of the neuronal networks necessary to support the brain's many functions. It is better to produce more neurons than are ultimately needed than to produce too few.

Neurons are not the only type of cell in the brain. Indeed, it turns out that another type of cell, the astrocyte, is as abundant as neurons. Astrocytes come from a different type of stem cell than do neurons. During brain development, neurons are born before astrocytes, so that during the first several months of gestation the human brain is mostly neurons. Thereafter, astrocytes are the main cell type added to the brain. Astrocytes have many important functions, including helping sustain energy levels in neurons, the metabolism of neurotransmitters, and the production of proteins that promote neuronal growth and synapse formation. Another type of brain cell is the oligodendrocyte. Oligodendrocytes wrap layers of their fatty membrane around the axon of a neuron in a process called "myelination." By insulating the axon in this manner, oligodendrocytes increase the velocity of the electrical impulse down the axon. Oligodendrocytes are abundant in the "white matter" of the brain, where bundles of long axons pass between brain regions on the same or opposite side of the brain or down to the spinal cord.

This chapter tells the story of the involvement of glutamate in the formation of neuronal networks during brain development. The questions I answer include: In what ways and how does glutamate affect the growth of dendrites and axons? How do synapses form, and what is glutamate's role in the process? What are the roles of glutamate in the production of neuronal growth factors and the death of some neurons during brain development? What do astrocytes do with glutamate that helps neighboring neurons? Does glutamate influence the myelination of axons? Understanding how the cellular architecture of the brain's neuronal networks is sculpted by glutamate during development has provided valuable insight into how this architecture is modified during adult life as well as its vulnerabilities during aging and disease.

SEEKING PARTNERS

As axons and dendrites grow in the developing brain, they probe their environment. To aid themselves in finding a neuron with which to form a synapse, they possess highly motile endings called "growth cones" (figure 3.1). Whereas the shaft of the axon or dendrite is cylindrical, the growth cone spreads out in and moves to and fro in a manner analogous to an amoeba. A typical growth cone is shaped somewhat like a hand, with fingerlike projections called "filopodia" that extend from the edges of the palm. In this analogy, the forearm would be the shaft of the axon or dendrite. The filopodia can elongate and retract and can attach and detach from molecules on the surfaces of other cells.

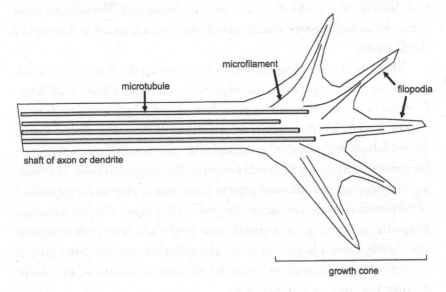

Figure 3.1

The basic structure of a growth cone. The growth cone is located at the distal end of a growing axon or dendrite. It is shaped like a hand, with a base or palm from which fingerlike structures called "filopodia" project. Two types of dynamic protein polymers control the growth of axons and dendrites. Microfilaments consisting of polymers of the protein actin control the extension or retraction of filopodia and their adhesion to the surface of other cells. Growth cones' interactions with other cells or the extracellular matrix determines whether an axon or dendrite turns or branches as it grows. Microtubules are constructed of tubulin polymers and control the lengthening or shortening of the axon or dendrite.

The elongation of the axon or dendrite and the shape and movements of growth cones are controlled by chainlike polymers of proteins. Within the shaft of the axon or dendrite are polymers of the protein tubulin. The tubulin polymers are organized so that they form hollow tubules with a diameter of about 25 nanometers. The filopodia of a growth cone do not have microtubules but instead have microfilaments that consist of polymers of the protein actin. Microfilaments have a diameter of about 6 nanometers. The lengths of microtubules and microfilaments can be increased or decreased. Microtubules have a polarity such that tubulin proteins are added to the end of a microtubule closest to the growth cone and removed from the other end. Therefore, the extension of an axon or dendrite requires that tubulin is added at the end of the microtubule nearest the growth cone, with little or no removal of tubulin from the other end. The same scenario is true for microfilaments, which extend when actin is added at their tips in the filopodia.

The growth cone is responsible for determining the direction in which an axon or dendrite grows, its rate of growth, and where a branch will form. Proteins in the membrane of a growth cone can attach and detach from anchor proteins on the surfaces of other cells or the extracellular matrix. Some of these "cell adhesion" proteins in the membrane of the growth cone are connected to the actin microfilaments inside the growth cone. The binding of a neuronal cell adhesion protein to an anchor protein on the surface of the growth cone can encourage the growth cone to grow in that direction. Filopodia on both sides of a growth cone might attach to anchor proteins and thereby cause a branch to form. The adhesion proteins that a growth cone encounters influence not only the direction of growth of an axon or dendrite but also how fast they grow.

Working with Viktor Hamburger at Washington University in Saint Louis in the early 1950s, Rita Levi-Montalcini observed that when she implanted cancer cells from mice into a chicken embryo growing within an egg, nerve cells in the developing chick grew much more quickly and grew toward the tumor cells. She then developed a cell culture system in which a small piece of tissue from the dorsal root ganglion is placed on the culture dish. Neurons in the dorsal root ganglion are sensory neurons. Their axons innervate the skin, muscle, tendons, and other tissues throughout the body. Levi-Montalcini and Hamburger found that a diffusible substance released by tumor cells and certain normal cells stimulated the growth of the axons of the cultured sensory neurons (Cowan 2001). Levi-Montalcini named the substance "nerve growth factor," or NGF. It was later determined that NGF is a protein.

Working at the University of Minnesota, Paul Letourneau and his colleagues found that growth cones of cultured sensory neurons move toward a focal source of NGF (Gallo, Lefcourt, and Letourneau 1997). When one side of the growth cone was exposed to NGF, the axon turned in that direction. They found that NGF causes the formation of filopodia as a result of increased polymerization of the microfilaments in the growth cone. NGF may also enhance the adhesion of the growth cone filopodia to proteins in the extracellular matrix.

Levi-Montalcini found that salivary glands produce very high amounts of NGF. This is very interesting from an evolutionary perspective because it is well known that many animals lick their wounds. If a mouse has a wound, other mice will even lick its wound for it. Such wound licking accelerates the closure of wounds in mice, and evidence suggests that NGF in the saliva is responsible for this beneficial effect of licking (A. Li et al. 1980).

Since the discovery of NGF, other proteins have been shown to promote the growth and survival of neurons. One such "neurotrophic factor" that is particularly important in the brain is the brain-derived neurotrophic factor, or BDNF. Proteins such as BDNF are encoded by genes on chromosomes in the nucleus. There are two copies of the BDNF gene. One is inherited from the mother and the other from the father. To determine if and how BDNF might influence the development of the brain, scientists used genetic-engineering technologies to disable or "knock out" one or both of the genes that encode BDNF in mice. When both BDNF genes are knocked out, the developing embryos die and have only a rudimentary brain. When one BDNF gene is knocked out such that the mice have a 50 percent reduction in BDNF levels, the mice survive but exhibit behavioral abnormalities, including impaired learning and memory and excessive anxiety. Examination of their brains revealed that there are reduced numbers of synapses in several regions of the brain and reduced neurogenesis in the hippocampus. Interestingly, the mice with reduced BDNF levels overeat and become obese because BDNF acts upon neurons in the hypothalamus to suppress appetite. Altogether, the evidence from mice shows that BDNF is essential for brain development and that a 50 percent reduction in BDNF levels adversely affects the development of neuronal networks, resulting in behavioral abnormalities (Vanevski and Xu 2013).

Cell adhesion molecules and neurotrophic factors help guide axons on their journey to find partner neurons with which they can form synapses. What about neurotransmitters? Do they have roles in brain development?

While working as a postdoctoral scientist at Colorado State University in the 1980s, I discovered that glutamate plays an important role in the formation of neuronal networks in the developing hippocampus. At the time I was performing my experiments, neuroscientists believed that glutamate acted only as a signal between neurons at synapses. To see whether glutamate might affect the outgrowth of dendrites or axons before synapses are formed, I cultured neurons taken from the hippocampus of developing rat embryos. During the first two days in culture, these neurons grew a long axon and several shorter dendrites (figure 1.2). After letting the neurons grow for several more days, I took pictures of them and then added glutamate to the culture medium. I then took pictures of the same neurons at 4, 8, and 24 hours after exposure to glutamate. For comparison, I took pictures of neurons in cultures that had not been exposed to glutamate.

My experiments were performed before the advent of digital cameras, so I used a 35-millimeter camera and black-and-white film. I would write down the sequence of pictures, develop the film, and then project the images of the neurons onto a piece of paper using a photographic enlarger used for printing. I traced the projected images of the neurons. I also took a picture of a "micro ruler" that I used to calculate the actual lengths of the axons and dendrites. All in all, a tedious process, but the results showed that glutamate inhibited the growth of the dendrites but had no effect on the growth of the axon (Mattson, Dou, and Kater 1988). When I looked at the growth cones of the dendrites under high magnification, I found that during the first several minutes after exposure to glutamate, filopodia extended. Then during the ensuing hours, the filopodia retracted, and the growth cone exhibited a rounded tip, much like the dendritic spine of a synapse. The questions then became: How does glutamate affect the growth of dendrites, and why does glutamate not affect the growth of axons?

In the mid-1980s, Roger Tsien developed fluorescent molecular probes that enable scientists to visualize various chemicals in living cells in real time, for which he was awarded the Nobel Prize in Chemistry in 2008. He named one of the molecular probes "Fura-2"; it enables the visualization of Ca^{2+} levels within cells (Grynkiewicz, Poenie, and Tsien 1985). Because electrophysiologists had previously shown that glutamate causes the influx of Ca^{2+} into neurons through channels in the membrane, I decided to use Fura-2 to see what happens to Ca^{2+} levels in the dendrites and axons when hippocampal neurons are exposed to glutamate. The Ca^{2+} levels increased rapidly in the dendrites but did not increase in the axon, presumably because the receptors for glutamate are located in the membrane of dendrites but not in the axon. The Ca^{2+} influx was responsible for the effects of glutamate on dendrite outgrowth because when a drug was used to block the Ca^{2+} influx, glutamate did not affect dendrite outgrowth (Mattson, Dou, and Kater 1988). The question then became: How does Ca^{2+} exert its effects on the growth of dendrites? Additional experiments showed that Ca^{2+} influx causes a rapid polymerization of microfilaments in filopodia and subsequently inhibits the polymerization of microtubules in the shaft of the dendrite.

Glutamate's ability to affect the growth of dendrites prompted me to ask whether it plays a role in the formation of neuronal circuits during brain development. To answer this question, I adopted a culture approach similar to that used by Rita Levi-Montalcini. I removed small pieces of tissue (explants) from a brain region called the "entorhinal cortex." Axons of glutamatergic neurons in the entorhinal cortex normally form synapses with neurons in the hippocampus. I asked whether glutamate released from those axons would affect the formation of synapses between the axons and hippocampal neurons (Mattson, Lee, et al. 1988). I first established entorhinal cortex explant cultures and let the axons of the neurons grow radially away from their cell bodies in the explant. I then placed hippocampal neurons in the cultures, some of which came to lie upon the axons of the neurons in the entorhinal cortex explant and some attached to the surface of the culture dish at a distance from the entorhinal cortex axons. I found that the dendrites of the hippocampal neurons growing on the entorhinal cortex neuron axons were shorter than those growing on the surface of the culture dish. To determine whether glutamate released from the axons was responsible for the reduced dendrite outgrowth, I treated the cultures with a drug that blocks glutamate receptors. This treatment resulted in an increase in the outgrowth of the dendrites, demonstrating that glutamate released from the axons was causing a reduction in dendrite outgrowth.

The next question was whether glutamate released from entorhinal cortex neuron axons affected the formation of synapses between those axons and the dendrites of hippocampal neurons. I used a staining method that enabled me to visualize and count synapses along the length of the dendrites. The results showed that glutamate released from axons causes synapses to form. Thus, glutamate not only acts at synapses but also plays a role in the establishment of those synapses during brain development. Hollis Cline at Yale University discovered that glutamate also plays critical roles in organization of neuronal circuits in the visual system of developing tadpoles (Cline and Constantine-Paton 1990). It is therefore likely that glutamate sculpts neuronal circuits throughout the nervous systems of all animals.

Soon after Cline and I published our discoveries demonstrating a critical role for glutamate in the formation of neuronal networks during brain development, Carla Shatz at Stanford University and others demonstrated the occurrence of spontaneous activity in neuronal circuits in rodents prior to the establishment of sensory inputs to the brain. This spontaneous activity is prevented with drugs that block glutamate receptors and prevent the formation of synaptic connections between neurons. A quote from an article published by Larry Katz and Carla Shatz in 1996 summarizes the conclusions of investigations of the mechanisms by which glutamatergic neuronal network activity constructs circuits in the cerebral cortex of laboratory animals: "Early in development, internally generated spontaneous activity sculpts circuits on the basis of the brain's 'best guess' at the initial configuration of connections necessary for function and survival. With the maturation of the sense organs, the developing brain relies less on spontaneous activity and increasingly on sensory experience. The sequential combination of spontaneously generated and experience-dependent neural activity endows that brain with an ongoing ability to accommodate to dynamically changing inputs during development and throughout life" (Katz and Shatz 1996, 1133).

Empirical evidence supporting glutamate's roles in the establishment of functional neuronal networks during brain development relied primarily on data from studies of mice and rats, but Peter Kirwan and his coworkers provided evidence that the same mechanisms occur in humans (Kirwan et al. 2015). They established cultures of human cerebral cortical neurons from pluripotent stem cells and showed that the neurons form large-scale networks with structural and functional characteristics similar to those of the developing human cerebral cortex. Glutamatergic neuronal networks with oscillatory activity patterns develop over several weeks. The frequency of the oscillations initially increases, then decreases, and finally exhibits organized activity patterns. The construction of these circuits and their ongoing spontaneous activity are abolished when glutamate receptors (AMPA and NMDA receptors) are blocked. It turns out that a small number of neurons receive more than four times the number of glutamatergic inputs that the rest of the neurons receive. Kirwan and the others believe that the neurons with more synapses represent hub neurons that coordinate ongoing patterns of activity in the cerebral cortex. Such activity patterns driven by glutamatergic synaptic activity may be the cellular basis of the electrical activity patterns seen in electroencephalogram (EEG) recordings made using electrodes placed on the scalp. It is thought that these oscillations are a mechanism that strengthens and fine-tunes synaptic connectivity throughout the brain.

Discovery of the mechanism by which glutamatergic neuronal networks are established during brain development raises a perplexing question. If Ca^{2+} influx is the signal that initiates the formation of excitatory glutamatergic synapses, then what is the signal for the formation of inhibitory GABAergic synapses? Yehezkel Ben-Ari and colleagues (2012) provided a possible answer. In the mature brain, GABA hyperpolarizes neurons and so does not increase but rather even decreases Ca^{2+} levels in the postsynaptic neuron.

The researchers used electrophysiological techniques to record responses of hippocampal pyramidal neurons to GABA in brain slice preparations from embryonic and postnatal rats. They found that GABA depolarized the pyramidal neurons in slices from embryonic and newborn rats. Then, during the first postnatal week, the response to GABA shifts from depolarization to hyperpolarization. Further experiments provided an explanation for this surprising result. The intracellular concentration of Cl– is higher in the embryonic neurons and then decreases as the GABA response shifts during early postnatal brain development. Therefore, during embryonic development, when GABAergic synapses are being established, GABA is depolarizing (i.e., it causes efflux of Cl– from the neuron) and so would be expected to cause Ca^{2+} influx. Calcium is therefore likely to be the intracellular signal that initiates the formation of GABAergic synapses during brain development.

A TIME TO DIE

You and I had many more neurons in our brains when we were in our mothers' wombs than we do now. In fact, it has been estimated that in some brain regions nearly half the neurons die before birth. Neuroscientists have found that there are two waves of neuronal death (Wong and Marin 2019). The first wave occurs soon after neurons arise from stem cells and before they grow axons and form synapses. The second wave of neuronal death occurs when their axons contact other neurons as they attempt to form synapses. In both instances, the neurons undergo a process of programmed cell death called "apoptosis." *Apoptosis* is a Greek word that refers to leaves falling off a tree. When a neuron dies by apoptosis, it shrinks while keeping its outer membrane intact. The dying neuron is then recognized by microglia, which are immune cells that gobble up the apoptotic neuron. The microglia can tell if a neuron is undergoing apoptosis because molecules on their outer membrane recognize a specific molecular change that occurs on the surface of the dying neuron.

An important feature of apoptosis is that it enables a cell to die without "spilling its guts." Thousands of cells in many tissues of the body undergo apoptosis every day. They are then replaced by new cells that arise from stem cells in those tissues. Epithelial cells such as those in the skin and those lining the intestines have a particularly high rate of apoptosis, which is matched by a high rate of production of new epithelial cells. This cell turnover ensures that older, "worn-out" cells are replaced by well-functioning young cells. But this is not true for neurons. They do not normally die during adult life and with few exceptions cannot be replaced if they do die.

Glutamate is thought to play an important role in determining which neurons live and which die in the second wave of cell death during brain development. All neurons have receptors for glutamate, and their activation results in Ca^{2+} influx into the cell. When a growing axon releases glutamate onto the dendrite of a potential target neuron, the resulting Ca^{2+} influx causes the cell to produce several neurotrophic factors. The neurotrophic factors are then released from the dendrite and activate receptors on the surface of the axon. Activation of the neurotrophic factor receptors, in turn, stimulates the production of proteins that prevent apoptosis (Burek and Oppenheim 1996).

Among the different neurotrophic factors that have been discovered, BDNF is the most robustly produced in the brain in response to activation of glutamate receptors. The increase in BDNF production resulting from glutamate receptor activation serves two major functions in the developing nervous system: it facilitates the formation and stabilization of synapses, and it supports the survival of the neurons. BDNF acts to increase the size of a synapse and the number of mitochondria associated with that synapse. In this way, BDNF may increase the local amount of energy available to support the continued maintenance and function of the synapse (A. Cheng, Wan, et al. 2012). BDNF may prevent neuronal apoptosis by increasing the production of the anti-apoptotic proteins such as B-cell lymphoma 2 (Bcl-2), antioxidant enzymes, and proteins that repair damaged DNA.

Glutamate's ability to promote the formation and maintenance of synapses occurs only with certain levels and patterns of synaptic activity. If there is too little activity, insufficient amounts of neurotrophic factors are produced, but too much activity may cause a synapse to be eliminated. In 1998, I coined the term *synaptic apoptosis* to describe the elimination of individual synapses. I discovered that changes that occur in the synapses in response to glutamate receptor activation are the same as those that had previously been shown to be involved in the process of apoptosis (Mattson, Keller, and Begley 1998). One change is the activation of enzymes called "caspases." These enzymes chew up many different proteins inside the cell without damaging cell membranes. A second change occurs in the outer membrane of the cells that signals microglia to engulf and "eat" the apoptotic cell. The change in the membrane involves the movement of a lipid molecule called "phosphatidylserine" from the inner part of the membrane to the outer surface of the membrane. Using a fluorescent molecular probe, I was able to show that phosphatidylserine is present on the surface of synapses when hippocampal neurons are exposed to stressful conditions, such as those that occur when neurons are deprived of neurotrophic factors or are exposed to excessive amounts of glutamate. This finding suggested that individual synapses that receive too little neurotrophic factors or too much glutamate might be removed by microglia.

Neurons that establish their connections within neuronal networks during brain development normally survive and function throughout an animal's life. The activation of glutamate receptors and consequent production of neurotrophic factors such as BDNF, fibroblast growth factor 2 (FGF2), and NGF keep the neurons alive and functioning well. However, as described in the next chapter, the structure of those neuronal networks is far from static. Dendrites and axons may grow or regress. New synapses may form, and others may be eliminated.

GLIAL CELLS HELP COMPLETE THE BRAIN-CONSTRUCTION PROJECT

In addition to neurons, the brain contains three types of glial cells—astrocytes, oligodendrocytes, and microglia (figure 3.2). Astrocytes are by far the most abundant type of glial cell. Indeed, it has been estimated that there are at least as many astrocytes as there are neurons in the human brain (von Bartheld, Bahney, and Herculano-Houzel 2016). Astrocytes are widely distributed throughout the brain and are abundant in both gray matter and

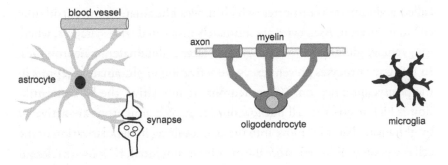

Figure 3.2

The three types of glial cells in the brain. *Astrocytes* are as numerous as neurons and serve many functions, including uptake of glutamate at synapses and coupling of neuronal activity with cerebral blood flow. *Oligodendrocytes* wrap around and thereby insulate axons in a process called "myelination." *Microglia* are immune cells that surveil the brain for pathogens and function in the pruning of unwanted synapses.

white matter. Oligodendrocytes are confined largely to white matter, where they wrap themselves around and thereby insulate axons. In this way, oligodendrocytes enhance the velocity of movement of an electrical impulse down the axon. Microglia are highly motile cells that function as part of the immune system. They surveil the brain in search of pathogenic microorganisms such as bacteria and viruses. They also gobble up cells and synapses that are undergoing apoptosis.

Until the mid-1800s, it was not known that we are made of individual cells. Then the German scientists Rudolf Virchow, Theodor Schwan, and Matthias Schleiden developed the "cell theory," which proposed that animals are made up of many individual living cells and that all cells come from other cells. Virchow's studies of the brain led him to conclude that glial cells function as a sort of "glue" that holds the brain together. It would be more than a century before the importance of glial cells in the formation and function of neuronal networks began to be appreciated. We now know that glial cells not only support the function and survival of neurons but also play roles in the sculpting of neuronal networks in the developing and adult brain.

One function of astrocytes is to remove glutamate and K^+ from the extracellular space. Within the outer membrane of astrocytes is a type of protein called a glutamate transporter, which moves glutamate from outside the cell to its interior. Astrocytes are intimately associated with synapses, where they remove glutamate from the synapse after a glutamatergic neuron fires. In this way astrocytes prevent excessive activation of glutamate receptors in the postsynaptic neuron. When neurons are not active, the K^+ concentration is high inside the cell but low outside it. When neurons are activated by glutamate, Na^+ ions move into the cell, resulting in depolarization of the cell membrane. To help restore the membrane potential, K^+ ions are released from the neuron into the extracellular space. For the neuron to recover from stimulation by glutamate, it is important that the K^+ be removed from the extracellular space. Studies have shown that astrocytes play a critical role in removing the K^+. If this function of astrocytes is compromised, neuronal networks become prone to hyperexcitability.

A second function of astrocytes is in the provision of energy to neurons. Astrocytes use a metabolic pathway called "glycolysis" to produce ATP from glucose. In contrast, neurons mainly use the mitochondrial electron transport chain—oxidative phosphorylation—to produce ATP from glucose. During the process of glycolysis, astrocytes produce lactic acid. The lactic acid is moved out of the astrocytes and into neurons, where it is used to produce ATP. Lactic acid also functions as a signaling molecule in neurons. By affecting gene expression, lactic acid increases the production of proteins involved in dendrite outgrowth and synapse formation in neurons. BDNF is one such protein.

Astrocytes produce several neurotrophic factors that are particularly important in supporting the function and survival of neurons. Two of them are FGF2 and insulin-like growth factor 1 (IGF1). Soon after I discovered that glutamate controls dendrite outgrowth and promotes synapse formation in the developing hippocampus, I asked whether neurotrophic factors might modify these effects of glutamate. Astrocytes produce and release FGF2. The FGF2 then binds to the surfaces of the astrocytes, where, I found, it stimulates the growth of dendrites and counteracts the inhibitory effect of glutamate on dendrite outgrowth (figure 3.3). When the growth cone of a growing axon or dendrite encounters an astrocyte, the FGF2 on the surface of the astrocyte stimulates growth along the surface of the astrocyte. In this

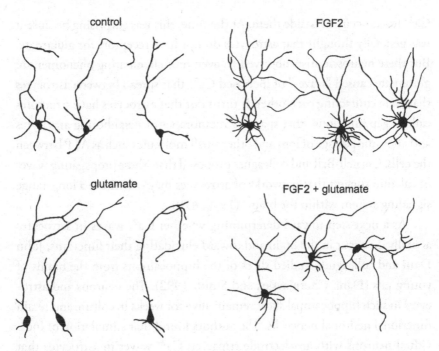

control   FGF2

glutamate   FGF2 + glutamate

Figure 3.3
Opposing effects of glutamate and FGF2 on the growth of embryonic hippocampal neurons. Glutamate inhibits the growth of dendrites without affecting the axon, whereas FGF2 enhances the growth of dendrites and axons. FGF2 counteracts the inhibitory effect of glutamate on dendrite growth. Adapted from figure 5 in Mattson, Murrain, et al. (1989).

way, astrocytes speed up the growth of axons and dendrites as they move toward the neurons with which they will form synapses. Calcium-imaging studies showed that this effect of FGF2 on neurons involves an attenuation of the elevation of intracellular Ca^{2+} levels caused by glutamate (Mattson, Murrain, et al. 1989).

An astrocyte typically resembles an octopus with many "arms" that reach out and encapsulate synapses. An individual astrocyte occupies a territory that partially overlaps with the territory of another astrocyte. Recent findings suggest that astrocyte networks may coordinate the activity of multiple synapses located along the dendrites of one neuron or even synapses located on different neurons. In 1990, Ann Cornell-Bell and Stephen Smith reported that when they exposed cultured astrocytes to glutamate, a rapid increase in

Ca^{2+} levels occurred inside them. At the time, this was surprising because it was generally thought that astrocytes do not have receptors for glutamate. But these neuroscientists observed an even more fascinating phenomenon: glutamate caused "waves" of increased Ca^{2+} that spread between astrocytes that were contacting each other. It turns out that astrocytes have structures called "gap junctions" that span the membranes of neighboring astrocytes and allow the passage of ions and other small molecules such as ATP between the cells. Cornell-Bell and colleagues proposed that "these propagating waves of calcium suggest that networks of astrocytes may constitute a long-range signaling system within the brain" (1990, 470).

As a next step toward determining whether Ca^{2+} waves in astrocytes actually do occur in the brain and toward elucidating their functions, John Dani and colleagues studied slices of the hippocampus from the brains of young rats (Dani, Chernjavsky, and Smith 1992). The neurons and astrocytes in such hippocampal slices remain alive for weeks in culture and retain functional neuronal networks. The authors found that stimulation of individual neurons with an electrode triggered Ca^{2+} waves in astrocytes that spread across the hippocampal slice. Further experiments provided evidence that astrocyte Ca^{2+} waves influence the activity of glutamatergic neurons. Because astrocytes arrive late in embryonic and early postnatal development, it is likely that Ca^{2+} waves play roles in the refinement of synaptic connectivity as nerve cell circuits are consolidated.

The type of glial cell that appears last in the developing brain is the oligodendrocyte. Oligodendrocytes arise from oligodendrocyte precursor cells, or OPCs. Oligodendrocytes wrap around the axons of neurons and thereby form an insulating sheath called "myelin." In humans, the last region of the brain where myelination of axons occurs is the prefrontal cortex. The nerve cell networks in the prefrontal cortex play important roles in making decisions, planning for the future, and moderating social behaviors. Myelination is not completed in this brain region until after adolescence. It is thought that teenagers' poor decision-making and their risk-taking behaviors are due at least in part to the fact that the neuronal networks in their prefrontal cortex have not yet matured with complete myelination.

Oligodendrocytes respond to glutamate in interesting ways. Dwight Bergles at Johns Hopkins University discovered that electrical stimulation of axons in the hippocampus results in the activation of glutamate receptors on OPCs (Bergles et al. 2000). When he looked closely at the OPCs using an electron microscope, he found that axon terminals form synapses with the OPCs. Glutamate apparently promotes the differentiation of the OPCs into oligodendrocytes, which can then myelinate axons (Gautier et al. 2015). Other studies have shown that exercise and maze learning—which involve increased activity in glutamatergic neuronal networks—can enhance myelination in rodents' brains (Tomlinson, Leiton, and Colognato 2016). However, social isolation results in reduced myelination in the prefrontal cortex. Evidence from animal studies suggests that the production of oligodendrocytes from OPCs is affected by early life experiences and that glutamate plays an important role in this process, although this remains to be established for humans.

Microglia are the brain's first line of defense against invading pathogens. They are the predators and garbage collectors of the immune system. The microglia perform their tasks of identifying and gobbling up microbes through a series of molecular interactions called the "complement cascade." Two key proteins in the complement cascade are C1q and C3. C1q is involved in recognition of the microbes, and C3 is involved in their destruction. But also beyond their role in destroying and removing pathogens, microglia function to fine-tune the structure of neuronal networks during brain development. Following up on my evidence that individual synapses can be eliminated by apoptosis, Beth Stevens and coworkers discovered a role for the complement cascade in the elimination of unwanted synapses in the developing brain (Stephan, Barres, and Stevens 2012). It was known that such "synaptic pruning" occurs in many brain regions during brain development. This pruning of synapses is thought to refine connections within neuronal networks so as to optimize their functions. It was known that C1q and C3 proteins were present in the brain, but it was generally assumed that they were involved only in pathological settings, such as an infection or brain injury.

To test the hypothesis that the complement cascade is involved in synaptic pruning in the healthy brain, Stevens and her colleagues studied the connections between neurons in the visual system of mice. Neurons in the eye called "retinal ganglion cells" send their axons through the optic nerve to the lateral geniculate nucleus, a region of the brain in the thalamus. This nucleus is a relay station that receives information from the retinal ganglion neurons and transfers it to neurons in the visual cortex at the back of the brain. Other neuroscientists had shown that the synapses in the lateral geniculate nucleus are arranged in columns, with one column dominated by axon terminals from the retinal ganglion cells of the left eye and the adjacent column dominated by axon terminals from retinal ganglion cells of the right eye. In mice, this segregation of synapses into ocular dominance columns occurs during the first two weeks after birth. If one eye is blinded, the segregation of synapses does not occur. The reason is that many synapses from the neurons of both eyes initially form, and then activity in each eye results in the elimination of some synapses and the strengthening of others. These synapses are glutamatergic, and so reciprocal interactions between glutamate and neurotrophic factors likely explain the formation of the ocular dominance columns.

Stevens and her lab members decided to see whether knocking out the gene for the C1q complement protein would affect the segregation of synapses in the lateral geniculate nucleus of mice. They recorded activation of synapses in the lateral geniculate nucleus in brain slices and visualized the synapses using antibodies for proteins located in synapses. They found that synaptic pruning was greatly reduced and that the segregation of synapses in the lateral geniculate nucleus did not occur in mice lacking C1q. Altogether, the evidence suggests that synapses destined for elimination are tagged with complement proteins. Microglia roaming the area detect the complement proteins and "gobble up" the unneeded synapses. Therefore, just as apoptosis eliminates neurons that do not integrate into neuronal networks during brain development, "synaptic apoptosis" eliminates individual synapses that are not necessary for the function of the neuronal networks.

In summary, astrocytes, oligodendrocytes, and microglia are increasingly recognized as playing important roles in the construction of nerve cell circuits in the brain during development. Astrocytes produce neurotrophic factors that enhance the growth of axons and the dendrites. They also remove glutamate and potassium after a neuron has been activated, which enables the neuron to recover its membrane potential so that it can be activated again. The oligodendrocytes myelinate axons, thereby enabling the axons to conduct impulses more rapidly. Microglia, which are the brain's immune cells, remove superfluous neurons and synapses, thereby contributing to the refinement of nerve cell networks. The activation of glutamatergic synapses affects all three types of glial cells, and vice versa.

4 FORGET ME NOT

Imagine that it were possible to build a brain from scratch, atom by atom, that has exactly the same molecular structure as yours. Would that brain encode all of the same memories and have the exact same personality, behavioral quirks, and beliefs that you do? In other words, would the synthetic brain hold your lifetime of experiences even though it did not actually experience them? If the answer is "yes," then memories, beliefs, and behavioral traits are determined entirely by the brain's molecular structure. If the answer is "no," then there must be another explanation for how memories are encoded in the brain.

"Let us assume," stated the Canadian psychologist Donald Hebb in 1949, "that the persistence or repetition of a reverberatory activity (or 'trace') tends to induce lasting cellular changes that add to its stability. . . . When an axon of cell A is near enough to excite a cell B and repeatedly or persistently takes part in firing it, some growth process or metabolic change takes place in one or both cells such that A's efficiency, as one of the cells firing B, is increased" (62), thus conceptualizing how the brain might encode memories. At the time, there were no actual experimental data to show that learning and memory involve the growth of a synapse or an enduring molecular change independent of growth. But Hebb's hunch proved valid.

Research on learning and memory is one of the largest subdisciplines of the neuroscience field. As such, many books have been written on various aspects of learning and memory, from the molecular and cellular level to the behavioral level. At present, there is no evidence of a mechanism for encoding memories other than one based on changes in the molecular organization of neuronal networks. Since Hebb's time, neuroscientists have made great strides in understanding some of the molecular changes involved in learning and memory. However, exactly how neuronal networks encode those memories remains a mystery. Nevertheless, it is clear that glutamate plays essential roles in learning and memory and orchestrates the structural and functional changes in neuronal networks that encode memories.

STRENGTHENING CONNECTIONS

A seminal discovery came in 1966 in the laboratory of Per Andersen at the University of Oslo, where Terje Lomo was studying how neurons in the hippocampus of rabbits respond to stimulation of the axons that innervate them. Lomo placed a stimulating electrode in the bundle of axons of neurons that innervate dentate granule neurons (figure 4.1) and a recording electrode in the dentate gyrus. He found that when he stimulated the axons at a high frequency (100 per second) for two seconds, the strength of the postsynaptic neurons' response to a subsequent single stimulus increased and remained elevated for more than an hour. Together with Tim Bliss, Lomo and Andersen further characterized this phenomenon, which is now called "long-term potentiation," or LTP (Lomo 2003). They found that LTP occurs only with certain frequencies of stimulation. For example, it occurs with a frequency of 100 stimulations per second but not with 10 stimulations per second. Subsequent studies showed that LTP also occurs in the hippocampus and other brain regions of living rats and mice. LTP is now considered to be fundamental to learning and memory.

The question then became what changes occur in the presynaptic and/or postsynaptic neurons that explain the strengthening of synaptic transmission in LTP? A clue came from the fact that both the presynaptic and postsynaptic neurons in Lomo's experiment are glutamatergic. Compelling evidence from many laboratories during the past four decades has shown that LTP involves the activation of AMPA receptors, membrane depolarization, and Ca^{2+} influx through NMDA receptor channels. It is also thought to involve

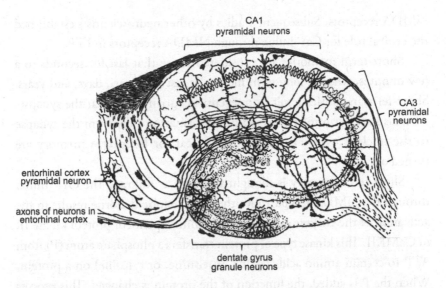

CA1
pyramidal neurons

CA3
pyramidal
neurons

entorhinal cortex
pyramidal neuron

axons of neurons in
entorhinal cortex

dentate gyrus
granule neurons

Figure 4.1
Drawing of neurons in the hippocampus by Santiago Ramón y Cajal in the early 1900s. Three main regions of the hippocampus are distinguished based on their location and connectivity— the dentate gyrus, region CA3, and region CA1. Information flows into the dentate gyrus of the hippocampus from pyramidal neurons in the entorhinal cortex. Dentate gyrus granule neurons project to CA3 pyramidal neurons, which in turn project to CA1 pyramidal neurons. All of these neurons—entorhinal pyramidal neurons, dentate gyrus granule neurons, and CA3 and CA1 pyramidal neurons—are glutamatergic.

the addition of new AMPA receptors into the postsynaptic membrane and an enlargement of the synapse.

Working at the University of British Columbia in the early 1980s, Graham Collingridge, Steven Kehl, and Hugh McLennan provided the first evidence that activation of NMDA receptors is necessary for LTP (Collingridge, Kehl, and McLennan 1983). They removed the hippocampus from adult rats and cut it into slices about one-third of a millimeter thick. An electrode was positioned to stimulate the axons of neurons that form glutamatergic synapses on the pyramidal neurons in region CA1 of the hippocampus. A recording electrode was placed so as to record the CA1 neurons' responses. Their experiments showed that LTP was abolished when they bathed the hippocampal slices with a solution containing a drug that selectively blocks

NMDA receptors. Subsequent studies by other neuroscientists established the critical role for Ca^{2+} influx through NMDA receptors in LTP.

Short-term memories are considered those that last for seconds to a few minutes, whereas long-term memories last for hours, days, and years. Short-term memories result from events occurring locally at the synapse, whereas long-term memories require information flow from the synapse to the nucleus, where certain genes critical for long-term memory are turned on.

Short-term memory is thought to occur as follows. The Ca^{2+} influx through the NMDA receptor into the postsynaptic dendrite results in the activation of the enzyme calcium/calmodulin-dependent protein kinase II, or CaMKII. This kinase type of protein transfers a phosphate atom (P) from ATP to certain amino acids (serine, threonine, or tyrosine) on a protein. When the P is added, the function of the protein is changed. This process is called "phosphorylation." The phosphorylation of several proteins at the synapse are involved in short-term memory. They include glutamate receptor proteins and proteins that control the microfilaments of the cytoskeleton. A particularly important effect of Ca^{2+} influx and CaMKII activation is the rapid insertion of glutamate receptors into the postsynaptic membrane of the dendrite. With more glutamate receptors in the membrane, the post-synaptic neuron's response to the glutamate released from the presynaptic terminal is increased.

In long-term memory, CaMKII activates a transcription factor called the "cyclic AMP response element-binding protein," or CREB. The activated CREB then moves to the neuron's nucleus, where it activates genes that encode proteins that are building blocks of synapses. In this way, the size of the synapse is increased. Moreover, long-term memories have been associated with increased numbers of synapses. Time-lapse imaging studies have shown that glutamate can induce the formation of filopodia from the shaft of a dendrite within a few minutes of activation of glutamate receptors and that these filopodia have the potential to become dendritic spines. Over the period of many hours, the filopodia can develop into spines. The spines may not initially be innervated by axons. Studies on the neocortex suggest that it takes several days for the spines to mature and become innervated and functional synapses. This sequence of events is very similar if not identical to the sequence of events that I discovered occurs during the initial formation of synapses in brain development (see chapter 3).

In addition to Ca^{2+}, a molecule called "cyclic AMP" plays an important role in learning and memory. The biochemist Earl Sutherland discovered cyclic AMP in the 1950s in his studies of how epinephrine (adrenaline) causes the mobilization of glucose from glycogen in the liver. He considered cyclic AMP to be a "second messenger" that transduces the signal from the first messenger epinephrine to elicit biochemical responses within the cell. Cyclic AMP activates protein kinase A, which phosphorylates many different proteins within the synapse. Protein kinase A also activates CREB.

Peter Guthrie was a postdoc in Ben Kater's laboratory during the time I was using the Ca^{2+}-sensitive probe Fura-2 to establish the importance of Ca^{2+} influx in the effects of glutamate on dendrite outgrowth and synapse formation. Guthrie filled a patch-clamp electrode with Fura-2 and injected it into individual CA1 pyramidal neurons in hippocampal slices. The Fura-2 diffused throughout the injected neuron, enabling the visualization of the dendrites and the spines thereon. When Guthrie exposed the neurons to a chemical that forms pores in the membrane that are permeable to Ca^{2+}, the concentration of Ca^{2+} throughout the dendrites increased, but there were clear differences in the Ca^{2+} concentration within individual dendritic spines (Guthrie, Segal, and Kater 1991). These findings demonstrated that individual spines differ in their abilities to reduce Ca^{2+} levels following its influx, which suggests that individual spines can differ in their responses to the same amount of glutamate receptor activation, which in turn suggests the possibility that the same amount of glutamate receptor activation may induce LTP at one synapse but not at another synapse.

One potential problem with the early studies of LTP is that the high-frequency stimulation most commonly used to induce LTP does not normally occur in neuronal networks in the brain. Electrodes placed into the hippocampus of freely behaving rats and EEG recordings of human brain activity have consistently shown electrical oscillations with frequencies of four to eight cycles per second. These oscillations are known as "theta rhythm." In rats, theta rhythm occurs when the animals are engaged in learning, ambulation, and exploratory sniffing. Theta rhythm also occurs during rapid eye movement (REM) sleep in both rats and humans. It is therefore thought to be involved in dreaming.

Theta rhythm stimulation can induce strong LTP at hippocampal synapses, and evidence suggests that theta rhythm plays a critical role in LTP and memory. It has been proposed that theta frequency is optimal for LTP and that it functions to separate the encoding of current sensory inputs from the retrieval of episodic memories related to the current sensory input. In this way, theta rhythms prevent the interference that would happen if the encoding of current sensory inputs and the retrieval of related memories occurred simultaneously.

Whereas patterns of synaptic activity such as theta rhythm can strengthen synapses, other patterns can weaken synapses in a process called "long-term depression," or LTD. It is thought that LTD plays an important role in the encoding of memories by preventing a synapse from reaching the ceiling level of potentiation. That is to say, if LTP at a synapse were allowed to reach a maximum level, then that synapse would no longer be able to encode new information. At glutamatergic synapses in the hippocampus, LTD occurs in response to persistent weak stimulation, whereas at the glutamatergic synapses on cerebellar Purkinje neurons LTD occurs in response to strong stimulation. Evidence suggests that LTD results from a reduction in the number of AMPA receptors in the postsynaptic dendritic spine (Ito 2001). As with LTP, LTD involves a Ca^{2+} influx through NMDA receptors. However, in contrast to LTP, where the Ca^{2+} activates kinases that phosphorylate certain proteins, in LTD the Ca^{2+} activates phosphatases that remove phosphate from the same proteins.

Most neurons in the brain have thousands of glutamatergic synapses. Whether a neuron fires an action potential in response to the activation of those synapses or not depends on how many synapses in proximity to each other are activated. Several theories have been proposed to explain how LTP or LTD at these synapses determines a neuron's ultimate response. One theory called "metaplasticity" posits a threshold of the postsynaptic neuron's response such that a response below the threshold results in LTD and a response above the threshold results in LTP (Abraham and Tate 1997). The threshold depends on the averaged activity of all the synapses on a neuron. Interestingly, LTD may have different effects on memory encoding depending on which neurons are involved. Early evidence suggested that LTD is involved in the decay of memory in hippocampal circuits but is critical for the learning of sequences of movements encoded in cerebellar circuits. However, subsequent studies suggested that LTD at hippocampal synapses is important for the encoding of memories related to spatial navigation. More recent findings suggest that LTP is important for encoding the position of one's body in a given space, such as forest or city, whereas LTD encodes objects and events encountered within the space.

Many people will also be shocked to learn that their brain cells produce gases that are potentially lethal! Neurons produce nitric oxide, hydrogen sulfide, and carbon monoxide when their glutamatergic synapses are activated, and evidence suggests that these gases are involved in learning and memory.

Most people have likely heard of instances of death caused by carbon monoxide poisoning from the burning of fuels in an enclosed space. Inhalation of high amounts of hydrogen sulfide—which smells like rotten eggs— can also be deadly. In fact, hydrogen sulfide gas was used as a chemical weapon in World War I. At levels of 100 parts per million or more, nitric oxide also poses an immediate danger to health and survival. All three gases are toxic at high concentrations, but brain cells produce them in very low amounts.

Except for carbon dioxide, the first gas found to be produced in animal cells was nitric oxide. In 1998, Ferid Murad, Robert Furchgott, and Louis Ignarro were awarded the Nobel Prize in Physiology or Medicine for their discoveries of "nitric oxide as a signaling molecule in the cardiovascular system" (Smith 1998, 1215). They discovered that nitric oxide is released from the endothelial cells that line arteries and then diffuses to the surrounding smooth-muscle cells and causes them to relax. In this way, nitric oxide increases the diameter of the artery, thereby increasing blood flow through the artery. This is why nitroglycerin is beneficial for patients with chest pain as a result of coronary artery disease: the nitroglycerin is converted to nitric oxide, which then relaxes the muscle cells surrounding the arteries and so increases blood flow and alleviates the pain.

Nitric oxide is also thought to have important functions in the brain. The neuroscientists David Bredt and Solomon Snyder (1990) at Johns Hopkins University discovered that nitric oxide is produced in neurons when glutamate receptors are activated. Its production results from Ca^{2+} influx. The Ca^{2+} binds to the protein calmodulin, which then causes the calmodulin to activate the enzyme nitric oxide synthase. This enzyme acts on the amino acid arginine in a manner that liberates nitric oxide. Nitric oxide then diffuses throughout the cell and activates the enzyme guanylate cyclase, resulting in the production of cyclic guanosine monophosphate, or "cyclic GMP." Similar to cyclic AMP, cyclic GMP activates a kinase enzyme that transfers the phosphate group from the GMP to proteins, such as ion channels and transcription factors.

Working at Stanford University, Erin Schuman and David Madison (1991) provided evidence that the nitric oxide produced in a postsynaptic neuron in response to glutamate receptor activation diffuses into the presynaptic axon, where it causes changes that affect glutamate release. Many other studies have since shown that nitric oxide functions as an intercellular signal that influences the structure and function of not only the activated synapse but also adjacent synapses. This is a unique feature of "gasotransmitters" such as nitric oxide. In contrast to neurotransmitters such as glutamate, which are constrained by membranes, the gases can freely diffuse across membranes.

The discovery that nitric oxide affects blood pressure and neuronal network activity via cyclic GMP led to the development of drugs that increase cyclic GMP levels for use in the clinic. For example, sildenafil (Viagra) and tadalafil (Cialis) are widely prescribed for men with erectile dysfunction. Sildenafil and tadalafil increase cyclic GMP levels in the muscle cells around the blood vessels in the penis, which relaxes them and so increases the amount of blood in the penis, with a corresponding increase in its size.

When Katsutoshi Furukawa was working in my laboratory at the University of Kentucky, he discovered that cyclic GMP reduces the excitability of ("relaxes") hippocampal neurons by activating K^+ channels in their membrane (Furukawa et al. 1996). This effect of cyclic GMP can prevent hyperexcitability and so protect the neurons from being damaged and killed by excessive glutamate receptor activation. Chapters 6 and 7 describe evidence suggesting that glutamate hyperexcitability occurs in a wide range of neurological disorders, including epilepsy, stroke, and Alzheimer's disease. Drugs that increase cyclic GMP levels are currently being evaluated for their potential benefit for patients with these disorders. Ottavio Arancio and coworkers at Columbia University found that sildenafil can improve learning and memory in a mouse model of Alzheimer's disease (Puzzo et al. 2009). Sildenafil increased the activation of the transcription factor CREB, suggesting that the drug activates genes that strengthen synapses. Other studies have shown that sildenafil can improve the outcome of traumatic brain injuries in animal models. The results of a small clinical trial of this drug in human patients who have suffered a traumatic brain injury suggest that sildenafil may improve their recovery (Kenney et al. 2018).

Nitric oxide is a free radical—that is, it has an unpaired electron and so has a propensity to "attack" proteins. Studies have shown that the amino acid tyrosine is particularly prone to attack by nitric oxide. When this occurs, the nitrogen from nitric oxide becomes tightly attached to the tyrosine in a process called "nitration." Depending on the amount of nitric oxide produced in response to glutamate receptor activation, protein nitration may play either a physiological role or a pathological role in synaptic plasticity. The relatively low amounts of nitric oxide production and protein nitration during theta rhythm activity may serve to modulate synaptic plasticity. However, excessive production of nitric oxide and protein nitration that occurs in neurons following an epileptic seizure or stroke may damage neurons.

Following up on his findings concerning nitric oxide, Solomon Snyder asked whether other gases might also function as messengers in the brain. He and his colleagues discovered that carbon monoxide and hydrogen sulfide are produced in brain cells. Like nitric oxide, both carbon monoxide and hydrogen sulfide are produced when synapses are activated by glutamate (Mustafa, Gadalla, and Snyder 2009; Paul and Snyder 2018). Carbon monoxide is produced when the enzyme heme oxygenase 2 acts on the iron-binding protein heme. Heme is best known as the protein that holds iron in the hemoglobin of red blood cells. But heme is also present in other cells, including neurons. The Ca^{2+} influx resulting from glutamate receptor activation stimulates heme oxygenase 2 and thereby causes carbon monoxide production. Like nitric oxide, carbon monoxide diffuses to neighboring neurons, where it increases cyclic GMP levels and activates K^+ channels. In this way, carbon monoxide can influence neuronal network activity.

Hydrogen sulfide was present on Earth long before life emerged. It is present in volcanic gases, hot springs, and natural gas. Cells now use the amino acid cysteine as a source of the sulfur to produce hydrogen sulfide. The production of hydrogen sulfide occurs by enzymatic reactions that are facilitated by vitamins B6 and B12 and folic acid. Upon its production, hydrogen sulfide diffuses within and outside of the cell in which it is produced. Although research on the functions of hydrogen sulfide in the brain is in its infancy, it is known that its sulfur atom can bind to the amino acid cysteine present in many proteins in a process called "sulfhydration." It is not yet known how many proteins are affected in this way by hydrogen sulfide or if their sulfhydration affects learning and memory. Nevertheless, evidence from electrophysiological studies of mouse hippocampal neurons suggests that hydrogen sulfide may play a role in synaptic plasticity (Abe and Kimura 1996).

THE BRAIN'S LEARNING AND MEMORY HUB

Perhaps the most important function of the brain is to enable an organism to learn, remember, and recall the details of its surroundings and experiences as it moves through its environment. Survival depends on the ability to remember the location of food sources, hazards, enemies, and friends. Assuming that we are not focused on our cell phone, our brain encodes a "cognitive map" of our environment. We do not remember every detail of every object during our journey. Instead, we remember key landmarks—a stream crossing, a rock formation, a street sign, or a restaurant.

The hippocampus is particularly important for such spatial navigation. There are two of them, one on each side of the brain, and they are the brain's learning and memory hubs. The word *hippocampus* comes from the ancient Greek word for "seahorse," *hippokampos*—*hippos* means "horse," and *kampos* means "sea monster." The human hippocampus is shaped like a seahorse and is located deep beneath the cerebral cortex. Mammals, birds, and lizards all have a hippocampus. Although mice, geese, and iguanas may not have the higher cognitive capabilities of humans, they can navigate very well.

Studies of humans who have suffered damage to the hippocampus and of laboratory animals whose hippocampus has been experimentally disabled have established the essential role of the hippocampus in learning and memory. The story of a young man named Henry Molaison is illustrative (Augustinack et al. 2014). Molaison was suffering greatly from epileptic seizures. In 1953, surgery was performed to remove the brain tissue that was thought to be where the seizures originated. Two-thirds of each hippocampus on the left and right were removed. The surgery was successful in reducing Molaison's seizures, but there was a major complication—he was no longer able to form new memories. He had great difficulty learning and remembering new routes and so was largely dependent on others. Interestingly, however, he did learn to navigate in a new house that he moved into five years after the surgery and could even draw a map of the room layout. It is thought that this may have been due to the facts that not all of each hippocampus was removed and that other brain regions were able to compensate for the loss of most of the hippocampus.

Instances of brain damage caused by trauma have also provided valuable information on the functions of different brain regions. However, such patients often have damage to more than one brain region. More precise interrogation of the relationships between brain structure and function have come from controlled lesioning of specific areas in the brains of laboratory rodents or monkeys. Lesions can be done by surgical resection, but even more precise lesions can be accomplished by focal injection of toxins that destroy neurons through excessive activation of glutamate receptors. Chapter 6 describes the discovery and uses of several such "excitotoxins," including kainic acid, ibotenic acid, and domoic acid. In the administration of an excitotoxin, the animal is anesthetized and then positioned in a stereotaxic apparatus that holds its head steady and enables precise positioning of the tip of an injection needle. A small hole is drilled in the skull, the needle is lowered to the injection site, and a few nanoliters of saline containing a high concentration of the excitotoxin is injected.

Such brain lesion studies in animals have established the importance of the hippocampus and closely associated brain regions with learning and remembering complex routes. Rats and mice with excitotoxic lesions that kill hippocampal pyramidal neurons are unable to learn to navigate through mazes. But this is only true if both the left and right hippocampi are lesioned. The latter fact suggests that the brain evolved as a bilateral organ in part as a means of redundancy so that if one side fails, the other side can carry on.

More than 100 years ago, the Spanish neuroanatomist Santiago Ramón y Cajal used a new tissue-staining method to reveal the remarkable structural organization of neuronal networks within the hippocampus. The staining method was developed by Camilo Golgi and enabled visualization of individual neurons. Ramón y Cajal made slices of the hippocampus along its length, much as one would slice a cucumber. He stained the slices with Golgi's stain and then made illustrations of what he saw under a microscope. An example of one his drawings is shown in figure 4.1. Cajal saw two C-shaped rows of neurons, one with large pyramid-shaped cell bodies and the other with smaller, rounder cell bodies. These rows are now called the "pyramidal cell layer" and the "dentate gyrus," respectively. Ramón y Cajal correctly surmised that information flows into the dentate gyrus from neurons in an adjacent region, the entorhinal cortex. Thence, the information flows from neurons in the dentate gyrus to neurons in the pyramidal cell layer. The axons of pyramidal neurons project to many different brain regions, including the prefrontal cortex, the amygdala, and the hypothalamus. We now know that all of the dentate granule neurons and all of the pyramidal neurons deploy glutamate as their neurotransmitter.

Essentially all information coming into the brain finds its way to the hippocampi. For example, in the case of the visual system, the pathway to each hippocampus is: retinal ganglion cells in the eye lateral geniculate nucleus in the thalamus primary visual cortex visual association cortex entorhinal cortex hippocampus. All of the neurons in this and other sensory pathways use glutamate as their neurotransmitter. Because inputs from all our senses ultimately reach the hippocampi, the latter can be considered hubs for learning and remembering the associations between different

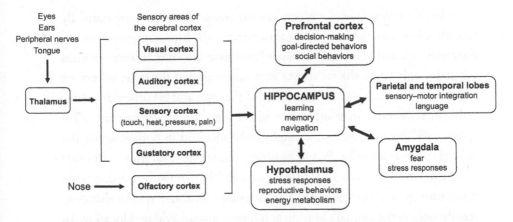

Figure 4.2

Neural pathways involved in the acquisition and processing of sensory information. Sensory neurons in the eyes, ears, tongue, and spinal cord (peripheral nerves) synapse upon glutamatergic neurons in the thalamus. The thalamic neurons project to corresponding sensory regions of the cerebral cortex. Olfactory inputs to the cerebral cortex bypass the thalamus. Information from all sensory organs "funnels" into the hippocampi, where glutamatergic neuronal circuits play critical roles in learning and memory, spatial navigation, and the construction of cognitive maps. Via glutamatergic pathways, the hippocampi have reciprocal connections with the prefrontal cortex, parietal and temporal lobes, amygdala, and hypothalamus. Some of the functions of these different brain regions are shown.

sensory inputs (figure 4.2)—for instance, associating a growl with dog, a foul smell with rotten food, a certain taste with vanilla ice cream, or a voice with a specific person.

Imagine that you are a hunter-gatherer living just north of what is now known as Lake Victoria in Uganda 20,000 years ago. Each day you and several others in your tribe gather your bows and arrows and walk through the forests stalking large and small game animals. You have learned to track animals based on their footprints and other clues. You have also learned that the animals have daily routines and so are in certain locations at certain times of day. Each time you venture out on the hunt, you and your cohunters are building and updating a "cognitive map" in the neuronal networks of your brains. This cognitive map enables you to remember the locations of water holes, animal dens, fruit trees, and so on. The hippocampus is where such cognitive maps are established.

How do neurons in the hippocampus enable accurate navigation? By recording the activities of individual neurons in the hippocampus in rats as they move around in a cage, neuroscientists have shown that some neurons are active only when the rat moves into one corner of the cage; others are active only when the rat is in the middle of the cage; and yet others are active only when the rat is in another area of the cage. These neurons are called "place cells" because they are thought to be involved in remembering the animal's current location. But recognizing one's current location is not sufficient to navigate toward a desired location. There must be neurons that function as part of something like a Cartesian coordinate system that constantly assesses the animal's bearing as it moves toward a desired location. In 2005, neurons located in the entorhinal cortex were discovered that fire at regular intervals as an animal navigates. Such neurons are called "grid cells" and enable the animal to know its current position in an environment by integrating information about the direction and distance of other locations with its current location. Within the entorhinal cortex, each grid cell is located at an equal distance from its neighbors.

The discoveries of hippocampal place cells and entorhinal grid cells were major advances in understanding how individual neurons work together to build a dynamic cognitive map. Place cells were discovered by the US researcher John O'Keefe, and grid cells were discovered by the Norwegian scientists May-Britt Moser and Edvard Moser (Hartley, Burgess, and O'Keefe 2013; Moser, Rowland, and Moser 2015). Because of their contributions to understanding how the brain works, these three neuroscientists were awarded the Nobel Prize in Physiology or Medicine in 2014. Since then, James Knierim at Johns Hopkins University has elucidated the cellular basis of "cognitive maps" by showing that hippocampal place cells "measure" the distance and direction traveled from a previous location. He used an "augmented-reality system" in which landmarks in the rat's visual field were moved in unison with the rat's movement on a circular track (Jayakumar et al. 2019). When the movement of the landmarks was stopped, the place cells "recalibrated" the path to update the rat's position in relation to visual landmarks. This result suggests that place cells are involved in the rapid updating of cognitive maps. An example of the rapid updating of a cognitive map in humans is in soccer, where the positions of players on a person's team and of players on the opposing team on the field are constantly changing, and success therefore depends on modifying one's cognitive map accordingly.

The research on place and grid cells was done in rats using electrodes surgically implanted into their hippocampus and entorhinal cortex. It would be unethical to perform similar experiments in humans. However, neuroscientists have used noninvasive fMRI to see whether neuronal circuits in the hippocampus or entorhinal cortex are active when human subjects are "navigating" through virtual environments. In one study, the neuroscientists Aidan Horner and Neil Burgess at University College London performed fMRI in people as they were imagining that they were moving around in a space with objects that they had become familiar with prior to the imaging (Horner et al. 2016). Their data provide evidence that neurons in the entorhinal cortex fire in hexagonal patterns as you move through your imagined space. Such grid cells are believed to be involved in the dynamic computation of your position by continually updating information about location and direction. Lose your grid cells, and you lose your bearings.

Understanding place and grid cells may prove valuable in illuminating the causes of neurological disorders. In fact, there is evidence that some grid cells degenerate late in life, and it is known that a massive loss of these cells has occurred in Alzheimer's disease. This makes sense because one of the symptoms of Alzheimer's disease is an impaired ability to remember where you have gone and what you did there. Lose grid cells, and you will not be able to locate your car in the mall parking lot.

The size of each hippocampus can be accurately measured in living persons by analyzing brain scan images. One interesting study showed that the size of the hippocampus of London cab drivers is greater than that of Londoners who are not cabbies (Maguire et al. 2000). This makes sense: a major function of the hippocampus is in spatial navigation, and cab drivers must have detailed cognitive maps of London stored in their neuronal networks. During spatial navigation, there is increased activity of neurons that encode a person's current location and planned route. Activation of glutamate receptors in these neurons stimulates the strengthening of existing synapses and the formation of new synapses. As a result, the overall size of the hippocampus increases. Because the study of cab drivers was done prior to the use of cell phones with GPS, it will be interesting to see whether the hippocampus is diminished in size in current cab drivers who do not have to have a map in their brain compared to cab drivers of "olden days."

In cooperation with closely associated brain regions such as the entorhinal cortex and inferior parietal cortex, the hippocampus encodes the recognition of objects and their relationships to each other in three-dimensional space. Face recognition is one remarkable example of the brain's ability to quickly learn and recall the distinguishing features of a complex object. We meet a person once, and we are able to recognize them months later, although we may sometimes not be able to remember their name. Although the hippocampus is necessary for learning the features of a face, it is not sufficient for remembering and recalling the neural representation of the face. Functional MRI studies of humans and monkeys have identified a region called the "anterior face area" in the temporal lobe close to the hippocampus where facial identities are stored. However, if the hippocampus on each side of the brain is lesioned, then facial identities cannot be encoded in the anterior face area. This is a common theme for all types of memories—they are initially encoded in the hippocampus, but their long-term storage occurs in neuronal circuits of the cerebral cortex.

The hippocampus has strong reciprocal glutamatergic connections with the frontal lobe. Circuits in the frontal lobe play critical roles in working memory, episodic memory, executive function, and decision-making. Working memory is a cognitive system that holds information temporarily. It is a type of memory that enables the manipulation of the stored information. An example of working memory is using a recipe that has not been previously memorized to cook a dish. Doing this involves briefly retaining one or several steps of the recipe in short-term memory as you prepare the dish. Episodic memory, in contrast, is a type of long-term memory by which you remember details of events such as places, times, places, objects, and people. For instance, you remember in the evening where you were and what you saw and did that morning and throughout the day. Executive functions are cognitive processes by which we modify our behaviors in a goal-directed manner. They include reasoning, planning, and problem-solving.

There is also inferential reasoning. "Inference can be thought of as a logical process by which elements of individual existing memories are retrieved and recombined to answer novel questions. Such flexible retrieval is sub-served by the hippocampus and is thought to require specialized hippocampal encoding mechanisms that discretely code events such that event elements are individually accessible from memory. . . . [B]y recalling past events during new experiences, connections can be created between newly formed and existing memories" (Zeithamova, Schlichting, and Preson 2012, 70). For example, you recently moved into a new house in a new neighborhood. As you are walking through a park a couple of blocks away, you see a teenage girl playing with a golden retriever with a bright-red collar. Three days later you see a middle-aged man walking the same dog on the sidewalk in front of your house. The man and his dog enter the house on the corner of the block. You conclude that the teenage girl is the daughter of the man and lives in the same house. The next day you see the same girl with a middle-aged woman riding bicycles on a street two blocks away. By inferential reasoning, you conclude that the woman is the wife of the man who lives in the house on the corner of your block.

The process of inference requires that relations among elements that were previously unrelated within an event must be encoded. In the example in the preceding paragraph, those elements are: teenage girl—golden retriever; middle-aged man—golden retriever; teenage girl—middle-aged woman. In this instance of inferential reasoning, the three binary associations must first be recalled and then manipulated, recombined, and recoded. The hippocampus is the hub of the glutamatergic signaling that is critical for inferential reasoning. Those glutamatergic circuits build memories that are used to make future decisions.

The essence of what many neuronal networks in the brain do is to process patterns of neuronal activities that encode images and sounds. Language is perhaps the best example of the brain's ability to process incoming patterns in ways that allow those patterns to be accurately regurgitated and manipulated in ways that enable the generation of new patterns. As I note elsewhere, "Rapid advances in science, technology and medicine are facilitated by language-based information transfer. However, despite it being a remarkable leap forward in evolution, language may not involve any fundamentally new cellular or molecular mechanisms; instead, language is mediated by recently evolved neural circuits integrated with older circuits, all of which utilize generic pattern processing mechanisms" (Mattson 2014, 265). Written languages are based on sequences of symbols that represent objects, emotions, actions, and so on. Spoken languages are based on sequences of sounds that correspond to the sequences of symbols in written language.

Language involves the working together of many different brain regions. Sensory pathways receive the images and sounds of letters and words and convey that information to the hippocampus. The patterns of the images and sounds are encoded in the hippocampus and then conveyed to regions of the cerebral cortex that integrate those patterns. Neuronal circuits in those brain regions process the information in ways that deal with various attributes of those patterns, including their meaning, salience, and emotional value. The neural-network-based representations of sequences of symbols and sounds are stored in short- and long-term memories. Particularly important for language are Broca's area in the lower part of the frontal lobe and Wernicke's area in the adjacent temporal lobe on the left side of the brain. Broca's area is critical for speech and so communicates closely with the motor cortex. Wernicke's area functions in the comprehension of language and for accurate communication by speech or writing. Neuronal circuits in these brain regions communicate via glutamatergic neurons with the circuits in the prefrontal cortex in the processes of creativity and decision-making.

In addition to being connected to brain regions involved in the long-term storage and recall of information, the hippocampus is connected to regions that apply emotional context to that information. One such region is the amygdala. Best known for its importance in memories of fearful experiences, the amygdala is a relatively small structure located immediately in front of the hippocampus. During a life-threatening or other scary situation, the initial memory of the particulars of the situation is encoded in the hippocampus and transferred to circuits in the amygdala, which creates the contextual association of the fear with the memory. Those particulars may include a location, a sound, an object such as a gun or knife, or another individual in distress. This communication between the hippocampus and amygdala evolved to increase animals' chances surviving in the wild—avoiding predators, for example. But repeated exposure to fearful situations can result in a debilitating repeated recall of those situations in PTSD. A fearful situation results in vigorous activation of glutamatergic neurons in the amygdala by glutamatergic neurons in the hippocampus. Fear memories result from activation of glutamate receptors on amygdala neurons, consequent influx of Ca^{2+} into those neurons, and activation of transcription factors such as CREB. By these mechanisms, the size and strength of "fear synapses" in the amygdala are increased.

The hippocampus also has important influences on the hypothalamus, a brain region that controls the production of hormones that regulate fertility, responses to stressful situations, growth, body temperature, water balance, and lactation (figure 4.3). Neurons in the hypothalamus produce small protein or "peptide" hormones that either directly or indirectly affect body functions. Some hypothalamic neurons produce the gonadotropin-releasing hormone (GnRH), which is released into the blood and acts locally on cells in the adjacent anterior pituitary gland, causing them to release luteinizing hormone (LH) and follicle-stimulating hormone (FSH) into the blood. The LH and FSH act on cells in the ovaries to increase estrogen production or in the testes to increase testosterone production. Other neurons in the hypothalamus produce corticotropin-releasing hormone (CRH), which stimulates the production of adrenocorticotropic hormone (ACTH). ACTH circulates in the blood and stimulates the production of cortisol in cells of the adrenal gland. Neurons in the hypothalamus also stimulate the production of pituitary growth hormone (GH), which then stimulates the production of IGF1 by liver cells. IGF1 promotes the growth of cells in many different tissues, including those in muscles and the brain. Hypothalamic neurons and pituitary cells also control the production of thyroid hormone, which regulates energy metabolism and body temperature. Finally, large secretory neurons in the hypothalamus release vasopressin or oxytocin. Vasopressin acts on cells in the kidney and thereby promotes water retention, while oxytocin acts on cells in the mammary glands to stimulate milk production.

The hippocampus has a major role in controlling the release of the hypothalamic peptide hormones. This has been most thoroughly studied

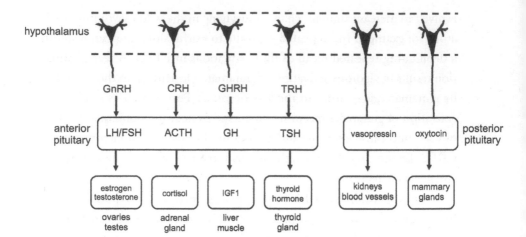

Figure 4.3

The hypothalamus–pituitary neuroendocrine system. Neurons in the hypothalamus project axons to the pituitary gland. The pituitary gland is divided into anterior and posterior regions. The anterior pituitary houses secretory cells that respond to peptides released from the axon terminals of hypothalamic neurons. Hypothalamic neurons can be divided into those that produce GnRH, CRH, GH-releasing hormone (GHRH), and thyrotropin-releasing hormone (TRH). GnRH stimulates pituitary cells that release leuteinizing hormone and follicle-stimulating hormone into the blood. CRH stimulates pituitary cells that release ACTH, while TRH stimulates pituitary cells that release thyroid-stimulating hormone (TSH). The posterior pituitary consists of the axon terminals of hypothalamic neurons that release the hormones vasopressin and oxytocin directly into the blood. Hormones released into the blood from the anterior or posterior pituitary exert effects on peripheral organs. LH and FSH stimulate the production and release of estrogen or testosterone from the ovaries or testes; ACTH stimulates release of cortisol from endocrine cells in the adrenal gland; GH stimulates the release of IGF1 from liver and muscle cells; and TSH stimulates the release of thyroxine from the thyroid gland. Vasopressin acts on cells in the kidney to promote water resorption and on vascular smooth muscle cells to cause vasoconstriction. Oxytocin acts on mammary glands to stimulate milk release, stimulates uterine contractions during childbirth, and is involved in social bonding.

in the case of the neurons that release CRH under stressful conditions. Glutamatergic neurons in the hippocampus activate GABAergic neurons in the hypothalamus, which in turn inhibit CRH production. In this way, the hippocampus is important for turning off the stress response after danger has passed. Evidence suggests that this important function of the hippocampus in turning off the stress response is impaired in PTSD and chronic-anxiety disorders. Chapter 9 considers the roles of glutamate in such disorders.

THE ELUSIVE ENGRAM

Electrical stimulation of certain regions of the cerebral cortex can elicit the recall of memories in humans. This was demonstrated in a series of studies performed by the neurosurgeon Wilder Penfield while he was working at the Montreal Neurological Institute in the 1930s and 1940s (Leblanc 2021). At that time, patients with severe epilepsy were often treated by surgically removing the brain tissue where the uncontrolled neuronal network excitability was thought to originate. The "focus" of the epileptic activity was most commonly in the temporal lobe and hippocampus. Before removing the brain tissue, Penfield placed a stimulating electrode in the patient's brain while the patient was still conscious and only under local anesthesia. This can be done because in contrast to other tissues throughout the body, the brain does not have pain receptors. Penfield found that stimulation of the temporal lobes resulted in the recall of memories that were often quite vivid. Some patients recalled not only memories of past events but also dreams or had visual or auditory hallucinations. Sometimes patients had déjà vu or "out-of-body" experiences. Penfield's findings provided evidence that the entire scope of human experiences is encoded in electrically excitable brain cells.

But how are the representations of visual, auditory, or other sensory information encoded in neuronal networks? An early theory came from Richard Semon, who was born in Berlin, Germany, in 1859. He was trained as an evolutionary biologist and later became fascinated with the question of how the memory of an experience persists in the brain and can be recalled later.

Semon coined the term *engram* to describe such a memory trace (Josselyn, Kohler, and Frankland 2017). He defined it as "the enduring though primarily latent modifications in the irritable substance produced by a stimulus" (Semon 1921, 12). At the time, neuroscience was a nascent field, and it was not yet known that the brain has networks of individual nerve cells that communicate via electrochemical transmission at synapses. Given this lack of knowledge, Semon would not speculate on the chemical nature of the engram and stated in a publication in 1923 that it is "a hopeless undertaking at the present stage of our knowledge and for my part, I renounce the task" (154). Although tremendous progress has been made since 1923 in understanding the cellular and molecular mechanisms of memory, the exact nature of the engram has remained elusive.

In the 1930s in the United States, the psychologist Karl Lashley performed experiments in rats that were designed to determine where in the brain engrams reside. He trained rats in a maze. He then removed tissue from different regions of the cerebral cortex and retested the rats in the maze. He found that the rats' ability to remember the maze was progressively diminished when increasing amounts of cerebral cortex were removed. But he also found that performance in the maze was impaired somewhat regardless of which region of the cerebral cortex was removed. These findings suggested that engrams are distributed throughout the cerebral cortex.

During the past 30 years, technological advances have accelerated progress in understanding the engram. Sheena Josselyn and Susumu Tonegawa describe four criteria that should be met for experiments aimed at establishing the nature of the engram: the data should (1) "show that the same or overlapping cell populations are activated both by an experience and by retrieval of that experience and should induce long-lasting modifications in those cells"; (2) demonstrate that "impairing engram cell function after an experience impairs subsequent memory retrieval"; (3) indicate that "artificially activating engram cells induces memory retrieval in the absence of any natural sensory retrieval cues"; (4) "artificially introduce an engram of an experience that never happened into the brain and show that rodents use the information of an artificial engram to guide behavior" (2020).

One technical advance in studies of memory mechanisms and the nature of the engram was the ability to identify neurons that were recently active by using an antibody to the protein called "Fos." Levels of Fos increase within minutes of stimulation of neurons and remain elevated for several hours. The Fos antibody will bind to Fos in those neurons. The location of the Fos antibody in brain tissue sections can then be visualized using immunohistochemistry, in which the Fos antibody is labeled with a fluorescent probe. For example, this method can be used in an experiment with two groups of rats. On the first day of the experiment, the rats in one group are trained to learn a maze, and the rats in the other group (the control group) are placed at the beginning of the maze but are not allowed to traverse the maze. The next day the rats in both groups are placed in one compartment in the middle of the maze for 10 minutes. The rats are then euthanized, and their brains are removed and processed for immunohistochemistry using the Fos antibody. Upon inspection of the brain tissue sections, it is clear that certain neurons in the hippocampus of the rats that had learned the maze are labeled with the Fos antibody but not in the hippocampus of the rats that did not learn the maze. This result is consistent with the possibility that the hippocampal neurons labeled with the Fos antibody are "engram cells" involved in the rats' memory of their location in the maze.

A major limitation of the Fos immunostaining method is that it cannot be applied to live animals. Therefore, it can provide only a snapshot of neuronal activity at the time the animal is euthanized. Ideally, neuroscientists would be able to observe the activities of individual neurons in real time in living and freely behaving animals. Two recent technological advances have enabled such studies and have revealed the ebb and flow of neuronal network activity in living mice. One advance was the development of fluorescent proteins that are responsive to Ca^{2+}. With genetic-engineering methods, the Ca^{2+} sensor protein can be introduced into all the neurons of mice or into specific subpopulations of neurons, such as GABAergic neurons or cholinergic neurons. Because activation of glutamate receptors is the trigger for Ca^{2+} influx into neurons, an increase in the fluorescence of the Ca^{2+} sensor protein is a measure of activation of glutamatergic synapses.

But imaging the Ca^{2+} sensor fluorescence in the brain of a living mouse is not easy. The first such studies involved removing a piece of skull from an anesthetized mouse and then putting the mouse under a microscope that was modified so that the objective lens could be positioned above the exposed region of the brain. A major limitation of this method is that only neurons located within 1 millimeter of the brain surface can be observed. To circumvent this limitation, neuroscientists developed a fiber optic technology in which a light-conducting fiber is inserted in the desired brain region, where the fluorescent Ca^{2+} sensor can then be imaged. Experiments using this method have advanced an understanding of the flow of information in neuronal networks and how glutamate and Ca^{2+} control that flow. For example, Japanese researchers used a fluorescent Ca^{2+} sensor called "yellow chameleon" to monitor the activities of neuronal networks in a brain region called the "parietal cortex." This brain region is involved in the integration of visual and auditory information. The researchers expressed the yellow chameleon protein in either glutamatergic or GABAergic neurons and found that spontaneous slow waves of Ca^{2+} self-organized into hubs with broadly active excitatory circuits and localized inhibitory circuits (Kuroki et al. 2018).

Until recently, the only way neuroscientists could study electrical activity in an individual neuron was by placing a recording microelectrode inside or on the membrane of the neuron. It was not possible, however, to record electrical activity in the many neurons of neuronal networks. For the latter, a voltage sensor that could be imaged at the microscopic level was needed. Ahmed Abdelfattah, Eric Schreiter, and their colleagues recently reported the development of a hybrid molecule consisting of parts of two proteins and a fluorescent molecule that intercalates into the membrane of cells and is highly sensitive to changes in the voltage across the membrane. They named the molecule "Voltron" (Abdelfattah et al. 2019). They used Voltron to image the activity of scores of neurons in the primary visual cortex in awake mice. When they moved a card with light and dark stripes in front of the mouse, some neurons fired axon potentials, whereas others did not. There were also clear patterns of correlated activity among neurons, suggesting that these patterns are neural representations of the visual patterns presented to the mice.

Voltron has also been applied in studies aimed at understanding how neuronal circuits interact to control body movements. For this application, scientists have taken advantage of the fact that the skin of the zebrafish is transparent, which makes it is possible to image cells throughout their body and brain without any invasive probes. Because the zebrafish is also amenable to genetic engineering, neuroscientists have used Voltron to study the activities of neurons in their brains. They monitored neuronal activity in a brain region that controls body movements during bouts of swimming induced by visual motion. They discovered neuron ensembles with activity patterns that were correlated with tail movements of the fish. There are neurons whose activity increases about one second before the fish starts swimming, neurons whose activity decreases every time the fish swims, and neurons that are continuously active during swimming.

A recent game-changer technical advance has elucidated the nature of the engram: optogenetics (C. Kim, Adhikar, and Deisseroth 2017). Optogenetics uses light to control the activity of neurons that have been genetically modified to produce ion channels that are sensitive to light. The most commonly used light-sensitive proteins are channelrhodopsin and halorhodopsin. Channelrhodopsin is an ion channel that fluxes Na^+ into the cell in response to light, whereas halorhodopsin moves Cl^- across the membrane into the cell. Shining a light on neurons with channelrhodopsin depolarizes the neuron in a manner similar to glutamate, whereas shining light on neurons with halorhodopsin inhibits the neuron in a manner similar to GABA. The methods were first worked out using cultured neurons, which can be directly imaged under a microscope and light shined on them while monitoring their activity with electrophysiology methods or by imaging Ca^{2+} levels in the neurons. Although the principle of optogenetics is relatively straightforward, applying it to living animals has been a major technological challenge. Transgenic lines of mice are first engineered so that the halorhodopsin or channelrhodopsin is expressed in all neurons or desired subtypes of neurons. Light-conducting glass fibers can then be placed in brain regions of interest. Shining light through the glass fiber will either excite or inhibit neurons in the vicinity of the fiber tip, and the light's effects on the behavior of the mouse can be determined.

Particularly important for engram studies was the development of transgenic mice in which neurons express only channelrhodopsin or only halorhodopsin after the neurons have been active. For example, glutamatergic neurons in the hippocampus that are active during maze learning can be "tagged" with either channelrhodposin or halorhodopsin.

One simple test of learning and memory in mice is fear conditioning, which can be done by pairing a sound with a foot shock, with the foot shock occurring about one second after the sound. This pairing is done several times on the first day of the test. On the second day, the mouse experiences the sound but not the foot shock. A mouse that remembers the association of the foot shock with the sound will freeze—it stops moving and tenses its muscles when it hears the sound. Using mice genetically engineered to express only channelrhodopsin in neurons activated during the fear learning, Xu Liu, Susumu Tonegawa, and their colleagues at the Massachusetts Institute of Technology were able to cause retrieval of the memory of a foot shock (X. Liu et al. 2012). The mice froze when the light was shined on certain neurons in the hippocampus. Their conclusion was that those neurons are "engram neurons" for that particular memory.

In an eerie advancement, neuroscientists have been able to implant false memories in mice (Ramirez et al. 2013) using optogenetic methods to stimulate either aversive (bad odor) or attractive (good odor) neural pathways in the olfactory system of the brain. A memory of an odor was implanted without the mouse having ever been exposed to that odor. From an evolutionary perspective, it makes sense that neuronal pathways that mediate responses to potentially dangerous exposures such as a bad odor or taste would be "hardwired." However, it seems less likely that false memories of specific individuals, objects, or events that have never previously been encountered can be physically (rather than psychologically) implanted in the brain.

Information processing by the brain exhibits features of both digital and analog systems. The all-or-none action potentials can be considered analogous to the binary digital system used for computer coding. But whether a neuron fires an action potential or not is determined by multiple upstream mechanisms that are continuously variable in nature. Individual neurons receive hundreds or thousands of synaptic inputs. And those synapses may be excitatory glutamatergic, inhibitory GABAergic, or modulatory in the cases of monoaminergic and cholinergic synapses. The summation of the activation state of these individual synapses determines whether a neuron fires an action potential or not. In addition, the timing of activation of excitatory, inhibitory, and modulatory synaptic inputs adds an additional level of control of neuronal activation. Moreover, the architecture of each neuron in the brain is unique and influences aspects of its electrical properties. Additional complexity comes from the dynamic nature of the structure of neuronal networks. Such neuroplasticity is thought to be important for the storage and recall of information. Beyond neurons themselves, glial cells are increasingly recognized for their roles in the control of neuronal excitability. For example, astrocytes rapidly remove glutamate from synapses, and oligodendrocytes accelerate the propagation of an impulse along an axon.

With so many levels of control of neuronal excitability in play, it is not surprising that it has been difficult to conceptualize how the brain works at a macrolevel. One theory of cognition posits that the brain operates by hierarchical extraction and parallel binding. Joe Tsien and colleagues at Princeton University describe this view of brain information processing: "These unique design principles enable the brain to extract commonalities through one or multiple exposures and to generate more abstract knowledge and generalized experiences. Such generalization and abstract representation of behavioral experiences have enabled humans and other animals to avoid the burden of remembering and storing each mnemonic detail. More importantly, by extracting the essential elements and abstract knowledge, animals can apply past experiences to future encounters that share essential features but vary in physical detail. These higher cognitive functions are obviously crucial for survival and reproduction of animal species" (L. Lin, Osan, and Tsien 2006, 54).

The neural mechanisms for memory encoding undoubtedly evolved to optimize success in survival and reproduction. But internal representations of external experiences need not involve the recording of the exact details of the external experience. Instead, the brain's neuronal networks extract features that are most important for successful adaptation. For example, imagine that you are hunting and see a deer. It is unimportant whether the deer has its ears pointed backward or forward or if the deer is facing to the left or to the right. What is important is that it is a deer and that it is close enough that your arrow will most likely hit it. Or imagine you are a surgeon performing a heart transplant. Your focus is entirely on the procedure.

Extraneous features in your visual field and sounds entering your auditory system are ignored. Only the neuronal networks in which the details of the sequence of surgical procedures and networks involved in real-time decision-making to adapt those memories to the features of the patient's thoracic anatomy are engaged.

Properly programmed computers can dramatically outperform human brains in their accuracy and ability to multitask. When endowed with rule-based algorithms, computers can also outperform humans in decision-making processes, as demonstrated by Deep Blue, which defeated the world's best chess player. But thus far computers have fallen far short of humans in their ability to generalize to new situations and to learn new things. In a recent review article on artificial intelligence, Fabian Sinz and colleagues described this fundamental problem in their field: "While rules in symbolic artificial intelligence provide a lot of structure for generalization in very narrowly defined tasks, we find ourselves unable to define rules for everyday tasks—tasks that seem trivial because biological intelligence performs them so effortlessly" (2019, 967).

A paradigm shift in artificial intelligence has resulted from the incorporation of analog computation approaches to machine learning, which involves multilayer artificial neural networks. For such "deep learning" of artificial intelligence algorithms, the individual artificial neurons summate multiple inputs from other neurons. Essentially, deep learning draws upon the fact that individual neurons have many excitatory glutamatergic synapses, whose activation can summate to reach a firing threshold. The integration into algorithms of the many other control systems operative in real neurons will be a daunting task for the artificial intelligence field if it desires to take that step.

Recent advances in the elucidation of the cellular and molecular nature of memory encoding and the functional organization of the brain's neuronal networks are exciting. However, they fall far short of a clear understanding of the molecular and cellular nature of the engram and how multiple engrams are organized to enable the storage and recall of sequences of individual memory traces.

5 SEEKING ENERGY

Success in the acquisition of energy and efficiency in its utilization for growth and survival is considered a fundamental driving force for the evolution of life. Humans store amounts of energy that enable them to live for extended periods of time—days, weeks, or even a few months—without eating anything. During such periods of food deprivation or fasting, the brain continues to function well as energy is prioritized for use by its neurons. In fact, when laboratory animals or humans are starved, all organs except the brain decrease in size. The ways in which the brain evolved to deal with variability in food availability are described in my book *The Intermittent Fasting Revolution* (2022).

An article entitled "Computers versus Brains" in the November 2011 issue of *Scientific American* (Fischetti 2020) caught my eye. The article compared the data-storage capacities, processing speed, and energy consumption of an iPad2, the fastest supercomputer, a human brain, and a cat brain. The data-storage capacity of the human brain was estimated to be a million times more than that of the iPad2 and only slightly less than that of the largest supercomputer. Even more remarkable was evidence that the processing speed of the human brain was similar to that of the supercomputer and much faster than that of the iPad. The cat brain was estimated to be similar to an iPad2 in its storage capacity and processing speed. The article also compared the energy efficiency of the human brain with that of the supercomputer and concluded that the human brain was about 500,000 times more energy efficient!

The brain consumes about 400 calories of energy during a 24-hour period. Therefore, the energy in three handfuls of almonds or a can of beans is sufficient to keep upward of 90 billion neurons and at least as many glial cells in the brain going strong for an entire day. The question then becomes: What are the mechanisms by which energy is distributed within and between brain cells? The short answer is that the mechanisms are based on simple principles but are complex with regard to their molecular and cellular dynamics. This chapter considers how glutamate plays a major role in brain "bioenergetics"—the distribution and utilization of energy by brain cells.

WHY NEURONS RESEMBLE TREES

Neuroscientists have often contemplated the similarities between the branches of a tree and the branches of a neuron's dendrites. Indeed, the terms *dendritic arbors* and *dendritic trees* are part of a neuroscientist's lexicon. I think that the structures of both tree branches and the dendrites of neurons are determined by molecular and cellular mechanisms that maximize energy acquisition and efficiency, which explains the similarity.

The branching patterns of trees and blood vessels exhibit what is called "fractal geometry." The Polish mathematician Benoit Mandelbrot coined the word *fractal* in developing a theory of self-similarity in nature. The structures of branches and blood vessels are considered fractal because they can be produced through simple recursive-branching rules (Glenny 2011). They begin as single trunks growing from the heart in the case of blood vessels and from the ground in the case of trees. Analysis of blood vessels in the human lung have shown that the pulmonary artery branches into approximately 70 million precapillary arterioles and 280 billion capillary segments. As the vessels branch, their diameter becomes progressively smaller. This pattern is similar to that of a tree arbor. Different species of trees have branching patterns that distinguish them from other species. Nevertheless, all of the branching patterns exhibit fractal geometry. Like trees, different types of neurons have distinct branching patterns. For example, the Purkinje neurons in the cerebellum have elaborate arbors that look very much like trees,

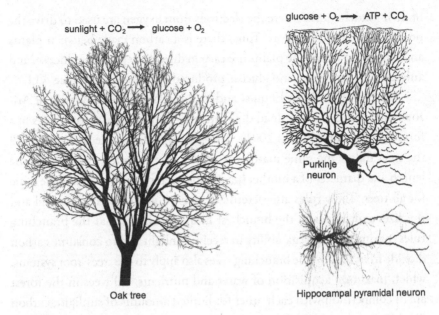

$$sunlight + CO_2 \longrightarrow glucose + O_2$$

$$glucose + O_2 \longrightarrow ATP + CO_2$$

Purkinje
neuron

Oak tree

Hippocampal pyramidal neuron

Figure 5.1

Examples of the branching patterns of an oak tree, a cerebellar Purkinje neuron, and a hippocampal pyramidal neuron. Cells in the leaves on the tips of branches use the energy in ultraviolet
sunlight to produce glucose from carbon dioxide in a process called "photosynthesis." Animal
cells use the energy in glucose plus oxygen to produce ATP in a process called "oxidative
phosphorylation." Photosynthesis results in the release of oxygen into the air, whereas oxidative
phosphorylation results in the release of carbon dioxide. This is the carbon cycle. The fractal
geometry of both tree branches and dendrites maximizes their abilities to acquire energy.

whereas pyramidal neurons in the hippocampus exhibit bushlike branches
that are fractal-like. (figure 5.1).

The branches of trees and other plants grow in a manner that maximizes their ability to acquire energy from sunlight. They grow toward areas
where sunlight exposure is greatest. This process is called "phototropism."
The leaves at the ends of the branches use the ultraviolet light from the sun
to produce glucose from carbon dioxide acquired from the air. Oxygen is
produced during this process. Cells throughout the plant use the glucose
produced in photosynthesis to produce ATP, which is the energy currency of
all cells. Plants can store glucose in complex carbohydrates and fats. Animals
that consume the plants utilize the carbohydrates and fats to produce ATP

in their mitochondria where the electrons from oxygen are used to drive the process of ATP production. Thus, there is a carbon cycle between plants and animals in which the plants use carbon dioxide to produce glucose, and animals use the oxygen and glucose produced by plants to produce ATP.

The biologists Brian Enquist and Karl Niklas at the University of Arizona meticulously determined the branching patterns of different trees in a forest (Enquist and Niklas 2001). They discovered that the branches of all the trees follow the same mathematical rule: for instance, the ratio of the length and diameter of a mother branch to its daughter branches is the same for all trees. These rules also determine where a new branch is formed and the sizes and shapes of the branches. Enquist believes that the branching rules maximize the plants' ability to receive sunlight and to consume carbon dioxide from the air. The branching rules also apply to the trees' root systems, which maximize acquisition of water and nutrients. All trees in the forest are in competition with each other for limited amounts of sunlight, carbon dioxide, water, and nutrients in the soil.

Chapter 3 described how the Ca^{2+} influx resulting from glutamate receptor activation sculpts the architecture of neurons during brain development. Plant biologists have recently discovered that glutamate also influences many aspects of plant morphogenesis. Much of this research was done using the thale cress plant, *Arabidopsis thaliana*. The plant biologist Brian Forde at Lancaster University found that application of glutamate to growing roots inhibits their elongation but at the same time stimulates branching of the roots (Forde 2014). What happens is that glutamate inhibits the division of cells in the root apical meristem while promoting lateral root initiation and growth. In this way, glutamate enables roots to maximize their surface area for the intake of water and nutrients. As is true for the effects of glutamate on dendrite outgrowth, Ca^{2+} mediates the effects of glutamate on root growth. The molecular biologist Myriam Charpentier and colleagues have further elucidated how Ca^{2+} controls the growth of plant roots. They showed that root growth is controlled by the movement of Ca^{2+} through ion channels in the nuclear membrane of the root cells: "Calcium (Ca^{2+}) is a universal regulatory element that intimately couples primary biotic and abiotic signals to many cellular process, allowing plants and animals to develop and adapt to environmental stimuli" (Leitao et al. 2019, 2). I certainly agree with Charpentier's statement.

I believe that in a way analogous to the maximization of acquisition of the sunlight by leaves on branches of plants, the growth cones and synapses on the ends of the branches of dendrites and axons seek to maximize their ability to acquire energy. But in the case of dendrites and axons, the acquisition of energy is indirect. Instead of sunlight, the dendrites and axons seek to maximize their acquisition of neurotrophic factors such as BDNF. The BDNF then activates receptors and signaling pathways inside the dendrite or axon that increase the number of mitochondria, which increases the local supply of ATP available to support the growth and branching of the axons and dendrites.

While working in my laboratory at the National Institute on Aging in Baltimore, the neuroscientist Aiwu Cheng discovered that an increase in the number of mitochondria in dendrites is required for BDNF to promote the formation of new synapses between hippocampal glutamatergic neurons. Her experiments drew upon previous research by scientists studying muscle cells who had shown that regular exercise increases the number of mitochondria in muscle cells by a process called "mitochondrial biogenesis," in which individual mitochondria divide and grow in size. It was discovered that the transcription factor peroxisome proliferator-activated receptor gamma coactivator alpha (PGC-1α) is activated in muscle cells in response to exercise, which results in the activation of genes that encode proteins critical for the division and growth of the mitochondria. Molecular biologists have identified several different genes induced by PGC-1α that encode proteins necessary for the growth of mitochondria and their division. These proteins include a protein called "translocase of the outer membrane" (TOM), which moves proteins into the mitochondria; mitochondrial transcription factor A (TFAM), which induces the expression of electron transport chain proteins; and the protein deacetylase situin-3 (SIRT3), which bolsters the resistance of mitochondria to stress.

To determine whether mitochondrial biogenesis is important for the formation and maintenance of synapses during development, Cheng first performed experiments with cultured hippocampal neurons (A. Cheng, Wan, et al. 2012). She used a fluorescent probe that labels mitochondria and showed that the number of mitochondria in the neurons increases dramatically as they grow and form synapses with other neurons. She then took advantage of a recent technology that can be used to selectively the block the production of any protein: RNA interference. The first step is to make a small synthetic RNA with a sequence complementary to a sequence in the messenger RNA that codes for the protein to be targeted. For Cheng's experiments, the small interfering RNA was targeted to the messenger RNA that codes for PGC-1α.

To introduce the small interfering RNA into cultured hippocampal neurons, Cheng took advantage of the fact that a virus called "adenovirus" can be genetically engineered to produce high amounts of such small interfering RNAs. The adenovirus readily infects neurons without having adverse effects on them. Cheng found that when PGC-1α levels were reduced, there was a decrease in mitochondrial biogenesis and a decrease in the number of glutamatergic synapses formed between the hippocampal neurons. We already knew that glutamate and BDNF play critical roles in the formation of synapses. As expected, Cheng found that when she exposed the cultured neurons to BDNF, the number of synapses increased. However, when she depleted PGC-1α from the neurons, BDNF was no longer able to promote synapse formation. In an additional experiment, the adenovirus with the PGC-1α small interfering RNA was injected into the hippocampus of adult mice. Over a period of several weeks, the number of synapses on the hippocampal neurons was decreased, suggesting that mitochondrial biogenesis was essential for the maintenance of synapses after they have been formed.

Emerging findings suggest the possibility that people can increase the number of mitochondria in neurons in their brains in three different ways—by engagement in intellectual challenges, by regular exercise, and by intermittent fasting. What these three ways of bolstering the ability of neurons to generate ATP have in common is that they are intermittent bioenergetic challenges. Chapter 11 is devoted to a discourse on how these three behavior/lifestyle modifications can improve the functional capabilities and resilience of glutamatergic and GABAergic neuronal circuits throughout the brain.

ASTROCYTES CONTROL ENERGY FLOW

GABAergic interneurons are intimately associated with glutamatergic neurons. It turns out that glutamate and GABA are produced from glucose. This is another link between neuronal energy metabolism and glutamate. The glutamatergic and GABAergic neurons depend on astrocytes for the production of their neurotransmitters. This process occurs as follows (figure 5.2): Astrocytes use glucose to produce the amino acid glutamine. The glutamine is then released from the astrocytes and transported into the adjacent neurons. Both glutamatergic and GABAergic neurons have an enzyme

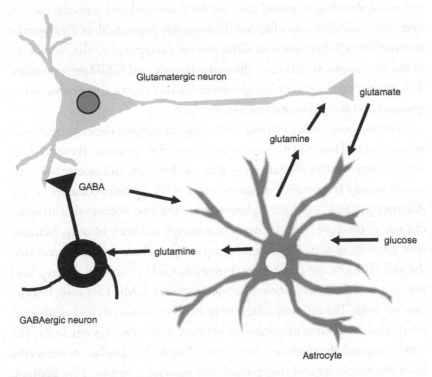

Figure 5.2
Astrocytes are critical for the production and metabolism of glutamate and GABA. Astrocytes produce glutamine from glucose. Glutamine is released from astrocytes and transported into neurons, where it is used to produce the neurotransmitters glutamate and GABA. Upon the release of glutamate and GABA at synapses, astrocytes take them up and convert them back into glutamine.

that converts glutamine to glutamate. The GABAergic neurons have an additional enzyme, glutamic acid decarboxylase, that converts glutamate to GABA. Once produced, glutamate and GABA are concentrated in the synaptic vesicles of presynaptic terminals. When they are released at the synapse, closely apposed astrocytes take up the glutamate and GABA and convert them back to glutamine.

The shuttling of glutamine from astrocytes to neurons and of glutamate and GABA from neurons to astrocytes provides an efficient recycling mechanism that ensures that neurons always have sufficient amounts of readily available glutamate or GABA for neurotransmission (Hertz 2013). This reciprocal shuttling of amino acids between neurons and astrocytes occurs over very small distances, often less than one ten-thousandth of a millimeter, because the cell membrane of astrocytes are juxtaposed to the membrane of the presynaptic terminals of the glutamatergic and GABAergic neurons. This means that the glutamine–glutamate/GABA shuttle speeds up at active synapses and slows down at inactive synapses.

Neurons require high amounts of energy to support their electrochemical activity, and much of that energy is consumed at synapses. There are three main energy sources for neurons—glucose, ketones, and lactate. Glucose comes mainly from carbohydrates in food, while ketones come from fats. Acetoacetate and beta-hydroxybutyrate are the two ketones that neurons can use to produce ATP, the molecular energy currency of cells. Neurons have proteins in their membranes that transport these energy sources into the cell. The glucose transporter 3 protein, GLUT3, moves glucose into neurons, and the monocarboxylic acid transporter 2, MCT2, moves ketones into neurons. The intimate relationship between astrocytes and neurons at synapses is designed to maximize the neurons' cellular energy efficiency. The astrocytes break down glucose into lactate, a molecule. The lactate is released from the astrocytes and transported into adjacent neurons. Like ketones, lactate is then moved into the mitochondria of the neurons, where it is used to produce ATP.

Because astrocytes are closely associated with synapses, the transfer of lactate from astrocytes to neurons is likely responsive to the energy demands of the adjacent synapses. However, we do not know this for sure because there is as yet no way to monitor the movement of lactate from astrocytes to neurons over such small distances.

The activity of neuronal networks consumes a great deal of energy. In fact, although the brain is only about 2 percent of the total body weight, it consumes up to 20 percent of the energy of a person at rest. The vast majority of the brain's energy consumption results from the activity of glutamatergic neurons. When activity in the neuronal networks of a particular brain region increases, the blood flow to that region increases to provide more oxygen and energy to the neurons. In fact, it is the blood flow that is being detected by fMRI brain scans and because more than 90 percent of the neurons throughout the brain are glutamatergic, fMRI signals can be considered a reflection of the activity of glutamatergic neurons.

It also turns out that astrocytes play a key role in the coupling of neuronal activity to cerebral blood flow. In the late nineteenth century, the Italian neuroscientist Camillo Golgi used a newly developed staining method on brain tissue sections and observed that astrocytes are intimately associated with cerebral blood vessels (figure 5.3). Individual astrocytes have extensive contacts with both synapses and cerebral blood vessels. Each astrocyte ensheathes several synapses and couples them with an adjacent blood vessel. When glutamate is released at the synapse, adjacent astrocytes respond with an elevation of intracellular Ca^{2+} levels. The Ca^{2+} stimulates the production of nitric oxide in the astrocytes. The nitric oxide then diffuses from the astrocytes to an adjacent blood vessel and causes the smooth muscle surrounding the vessel to relax, thereby increasing blood flow through that vessel. Thus, by acting on astrocytes, glutamate increases cerebral blood flow and thereby increases nutrient supply to neuronal networks that are active.

GLUTAMATE AND BDNF BOOST NEURONAL ENERGY PRODUCTION

The activity of glutamatergic synapses is the primary driver of energy demand throughout the brain. When neurons are more active, they require more energy to support that activity. Within seconds to minutes, activation of glutamate receptors causes a rapid increase in the production of ATP in

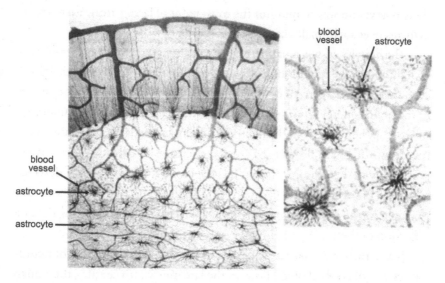

Figure 5.3

Astrocytes are intimately associated with cerebral blood vessels. The images are drawings made by Camillo Golgi and published in 1886 in his book *Sulla fina anatomia degli organi centrali del sistema nervoso* (table 12). The images are of sections of the human cerebellum that Golgi stained using a histological method he developed that strongly stains astrocytes. Golgi noted that blood vessels are often contacted by enlargements at the ends of astrocyte processes—structures that are now referred to as astrocyte "end feet."

the mitochondria located close to the synapse. The ways in which glutamate stimulates energy production in neurons are like the ways in which exercise increases the strength and endurance of muscles. Glutamate causes Ca^{2+} influx into neurons, and acetylcholine causes Ca^{2+} influx into muscle cells. In both neurons and muscle cells, the Ca^{2+} then rapidly moves into the mitochondria, where it stimulates the activity of the electron transport chain to increase ATP production.

Zheng Li, Morgan Sheng, and their colleagues at MIT showed that mitochondria move from the dendritic shaft to the base of the dendritic spines at glutamatergic synapses activated by electrical stimulation (Z. Li et al. 2014), which occurs within minutes of stimulation and requires the activation of NMDA receptors. Presumably, this rapid movement of mitochondria to active synapses increases the amount of ATP available to them.

Repeated activation of glutamate receptors as occurs in neuronal networks engaged during cognitive tasks or during physical activity causes long-term increases in those neurons' energy-producing ability. As with muscle cells, neurons that are more active require more energy to support their function. Accordingly, neurons being regularly activated by glutamate have greater numbers of mitochondria than do neurons that are less active (Raefsky and Mattson 2017).

The BDNF produced in response to glutamate receptor activation plays an important role in the ensuing mitochondrial biogenesis (Marosi and Mattson 2014). BDNF stimulates an increase in the number of mitochondria in neurons as they grow and form synapses during brain development. BDNF sets in motion a series of events that boost energy production in neurons. BDNF binds to the outer part of the receptor protein called "tyrosine protein kinase B" (TrkB) in the membrane of the cell (figure 5.4), which causes a change in the shape of the receptor on the inner side of the membrane, which in turn leads to the activation of several enzymes called "kinases." The kinases then activate the transcription factor CREB. Upon activation, CREB moves into the cell nucleus, where it activates many genes, some of which are involved in the bolstering of cellular energy levels. BDNF stimulates the transport of glucose into neurons and increases activity of the mitochondrial electron transport chain, thereby increasing ATP production.

Over time, mitochondria can become damaged and dysfunctional as a result of their high level of exposure to free radicals, which damage the DNA, proteins, and membranes in the mitochondria. One consequence of this damage is that the mitochondria become unable to maintain a normal protein gradient across their membrane. Cells have evolved an elaborate process by which dysfunctional mitochondria are identified and removed: mitophagy. A dysfunctional mitochondrion is moved into the lysosome, a membrane-bound compartment of the cell. The lysosome has a low acidic pH. When a dysfunctional mitochondrion enters the acid bath in the lysosome, its components are hydrolyzed by enzymes. The proteins in the mitochondria are broken down into amino acids, which can then be recycled and used to produce new proteins. Mitophagy is a specific type of autophagy, a more general cellular garbage-disposal and recycling system.

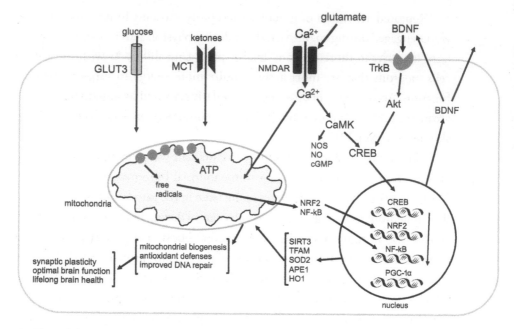

Figure 5.4

Glutamate affects mitochondria in ways that bolster their function and enhance neuroplasticity. Glutamate causes Ca^{2+} influx through NMDA receptors (NMDAR) and voltage-dependent Ca^{2+} channels. Calcium acts directly on mitochondria to increase ATP production, with an associated increase in the generation of superoxide and other free radicals. The increased free radical production results in the activation of the transcription factors nuclear factor erythroid 2–related factor 2 (NRF2) and nuclear factor kappa-light-chain-enhancer of activated B cells (NF-κB). By activating calcium/calmodulin-dependent protein kinase (CaMK), Ca^{2+} activates the transcription factor CREB. One gene activated by CREB encodes BDNF. BDNF is released and activates TrkB receptors on the same neuron or adjacent neurons, which results in the activation of the protein kinase B (Akt) and CREB. Via Ca^{2+}-mediated processes, including activation of CREB and the generation of free radicals, glutamate activates the transcription factor PGC-1α, which is a master regulator of mitochondrial biogenesis, the process by which cells increase the numbers of their mitochondria. Many of the genes induced by activation of glutamatergic synapses are involved in the bolstering of mitochondrial function and resilience. These include the protein deacetylase SIRT3, the mitochondrial transcription factor TFAM, the mitochondrial antioxidant enzymes superoxide dismutase 2 (SOD2) and heme oxygenase 1 (HO1), and the DNA repair enzyme AP endonuclease 1 (APE1). The beneficial effects of glutamatergic neuronal network activity on mitochondria are increasingly recognized as being critical for synaptic plasticity, optimal brain function, and lifelong brain health. cGMP = cyclic guanosine monophosphate; NO = nitric oxide; NOS = nitric oxide synthase.

Effects of BDNF on autophagy and mitophagy have been reported (Jin et al. 2019), but it is not clear if and how these processes affect the structure and function of neuronal networks. Nevertheless, it is known that when glutamatergic neurons are functioning properly, autophagy and mitophagy are highly efficient processes. It is also known that mitophagy is impaired in neurons affected by Alzheimer's and Parkinson's disease and that this abnormality contributes to the demise of the neurons in these diseases (see chapter 8). The accumulation of dysfunctional mitochondria causes an energy deficit in neurons and increases the production of free radicals. These abnormalities cause neurons to become hyperexcitable, resulting in aberrant neuronal network activity. The next chapter describes the process of excitotoxicity, in which neurons are damaged and killed as a result of being stimulated excessively by glutamate. In fact, such excitotoxicity is believed to occur in all the major brain disorders that manifest degeneration of neurons, including epilepsy, stroke, Alzheimer's disease, Parkinson's disease, ALS, and Huntington's disease.

During the process of ATP production in mitochondria, oxygen is consumed, and free radicals are produced. Simply put, an increase in energy production results in a concomitant increase in free radical production. As long as free radicals are dealt with efficiently, they do not adversely affect the cells. In fact, the opposite is true. The free radicals normally do good things for the muscle and nerve cells. They stimulate the activity of several transcription factors, which then set in motion several adaptive responses that ultimately make the cell more resistant to stress and better able to function. Two transcription factors activated by free radicals are nuclear factor erythroid 2–related factor 2 (NRF2) and nuclear factor kappa-light-chain-enhancer of activated B cells (NF-κB). These transcription factors induce the expression of genes that encode antioxidant enzymes such as superoxide dismutases, heme oxygenase, and glutathione peroxidase. In this way, exercising your muscle and brain cells makes them better able to cope with oxidative stress. This is a classic example of "hormesis," a beneficial adaptive response of cells and organisms to moderate intermittent stress.

Intellectual challenges and exercise are examples of moderate stressors that have beneficial effects on neurons. The cells are subjected to moderate metabolic and oxidative stress during their activation, and they respond adaptively by enhancing their plasticity (increasing synaptic strength and numbers of synapses) and their resistance to stress. Activation of glutamate receptors is thought to be involved in such adaptive responses. Indeed, 30 years ago Ann Marini discovered that exposure of cultured neurons to low concentrations of glutamate can protect them from being killed by the neurotoxin 1-methyl-4-phenylpyridinium (MPP^+) (Marini and Paul 1992). Evidence suggests that the BDNF produced in response to neuronal activity can enhance the neurons' intrinsic antioxidant defenses. For example, when cultured hippocampal neurons are exposed to BDNF, the neurons increase their production of several antioxidant enzymes (Mattson, Lovell, et al. 1995).

In addition to bolstering antioxidant defenses, the Ca^{2+} influx and free radical production caused by glutamate receptor activation may enhance neurons' ability to repair molecular damage caused by free radicals. Nucleic acids in DNA are often damaged by the free radicals produced during cell respiration. Because activation of glutamate receptors increases free radical production, it can also increase DNA damage. If not repaired, the damage can result in mutations in genes, and those mutations in turn can result in a change of the amino acid sequence of the protein encoded by the affected gene, which most commonly adversely affects the protein's function. Studies have shown that when neurons are stimulated by glutamate, their ability to repair DNA is improved. The Ca^{2+} influx and free radical production caused by glutamate result in a transient increase in DNA damage but at the same time set in motion signaling events that enhance the neuron's ability to repair that damage and future damage (J. Yang et al. 2010). One of those events is the increased production of BDNF. The BDNF induces the expression of an enzyme that removes damaged DNA bases, which are then replaced with pristine bases.

Chapter 4 pointed out that nitric oxide, which is a free radical, is produced when neurons are stimulated by glutamate and that the nitric oxide plays important roles in learning and memory and associated structural changes in synapses. Superoxide and hydrogen peroxide are two other reactive oxygen molecules produced in response to activation of glutamate receptors and involved in synaptic plasticity and cognition. Superoxide is a free radical produced in mitochondria during the electron transport process that generates ATP. Hydrogen peroxide is produced from superoxide. It is not a free radical but can cause the production of the hydroxyl radical, which if not rapidly squelched can damage cells. The neuroscientist Eric Klann and others have examined the effects of experimental manipulations of superoxide and hydrogen peroxide levels on LTP and cognition in mice. The take-home message from their work is that only a certain range of concentrations of these reactive molecules enables optimal synaptic plasticity (Serrano and Klann 2004). If there is too little or too much of these reactive molecules, LTP is reduced, and cognition is impaired. Those optimal amounts of superoxide and hydrogen peroxide are those that likely occur in response to the "exercising" of neuronal networks, as occurs when the individual is engaged in intellectually challenging endeavors.

6 EXCITED TO DEATH

Thus far we have seen how glutamate sculpts the formation of neuronal networks during brain development and how it modifies those networks adaptively in response to the demands placed on them during daily life. But glutamate also has the potential to destroy neurons. Excessive activation of glutamate receptors can outright kill neurons in the brain. The phenomenon of glutamate "excitotoxicity" can be directly observed by looking at cultured neurons exposed to a high concentration of glutamate. Such excitotoxic neuronal death is also evident upon examination of the brains of patients who have suffered from severe epileptic seizures and of laboratory animals in experimental models of epilepsy. But before I delve into what happens to neurons when they are excessively stimulated by glutamate, I have several interesting stories of natural accidents that illustrate the devastation that excitotoxicity can wreak upon the brain.

TALES OF SHELLFISH, SUGAR CANE, HEROIN USERS, AND PESTICIDES

As a result of the increase in atmospheric carbon dioxide, mostly from the burning of fossil fuels and deforestation by humans, our planet has been rapidly warming during the past five decades. Global warming is causing the melting of ice at the poles, rising sea levels, and the increased frequency and severities of hurricanes and forest fires. The frequency of harmful algal blooms, also known as "red tides," is also increasing.

In 1987, more than 100 people became ill after eating blue mussels at a restaurant in Prince Edward Island, Canada. Many of them experienced epileptic seizures and persistent memory loss, and three died. Analysis of the mussels revealed that they had a high concentration of the chemical domoic acid (Todd 1993). The domoic acid was not produced by the mussels but instead by the algae that the mussels consumed during a red-tide year. Domoic acid activates the kainic acid type of glutamate receptor. Whereas glutamate is rapidly removed from the extracellular space after it is released from presynaptic terminals, domoic acid is not removed. Therefore, domoic acid causes a continuous activation of glutamate receptors, which can cause seizures and the death of the neurons. Domoic acid is therefore considered to be an "excitotoxin." Pyramidal neurons in the hippocampus are particularly vulnerable to seizures and excitotoxicity because of their complement of glutamate receptors and the peculiarities of the neuronal connections in the hippocampus. The seizures and memory loss experienced by those who ate the mussels in Prince Edward Island resulted from the excitotoxic degeneration of neurons in their hippocampi.

In 1961, people in Santa Cruz, California, witnessed an invasion of crazed seabirds that attacked people. The birds were believed to have consumed shellfish containing high levels of domoic acid. This incident inspired a scene in the movie *The Birds*. Television shows such as *Elementary* have even included episodes where people were intentionally poisoned with domoic acid.

In addition to domoic acid, other excitotoxins are produced by various organisms throughout the world. In the 1950s, kainic acid was isolated from seaweed. It was discovered before the receptors for glutamate were identified, and it was used to establish the identity of the glutamate receptor that we now call the "kainic acid receptor." The kainic acid receptor fluxes high amounts of Na^+, causing depolarization of the neuronal membrane and influx of Ca^{2+} through the NMDA glutamate receptor channel. Another naturally occurring excitotoxin is ibotenic acid. It is produced by certain species of toxic mushrooms in the *Amanita* genus. As with domoic acid and kainic acid, ingestion of ibotenic acid can cause amnesia as a result of damage to hippocampal neurons. Other neuroscientists and I have used kainic acid and ibotenic acid to cause neuronal network hyperexcitability and excitotoxicity in rat and mouse models of epilepsy and Alzheimer's disease.

Whereas domoic acid, kainic acid, and ibotenic acid can directly activate glutamate receptors, other natural toxins cause excitotoxicity indirectly. Two such indirect excitotoxins have proven useful in our understanding of what happens in the brains of people with Huntington's disease and Parkinson's disease. The first story begins in northern China, where in the 1970s and 1980s more than 800 people, mostly children, were reported to have developed severe neurological symptoms, including seizures and coma. After recovering from the coma, they exhibited symptoms very similar to those of Huntington's disease: they were unable to control the movements of their limbs and so exhibited continuous body movements, or a phenomenon called "chorea." Investigations revealed that they all had eaten sugar cane. The sugar cane had been harvested in southern China and then shipped to northern China, where it was stored for several months. It turns out that the sugar cane was moldy and that a neurotoxin produced by the mold cells was responsible for the damage to the brains of the people who consumed the sugar cane. Researchers identified the culprit in the mold as 3-nitropropionic acid (3-NPA) (Brouillet et al. 2005).

Examination of the brains of some of the moldy sugar cane victims revealed extensive damage to the striatum, the brain region that is ravaged in Huntington's disease. To understand how degeneration of neurons in the striatum results in the chorea of people with Huntington's disease, it is necessary to know how the brain controls movements of the body. Movements are initiated by the activity of large neurons located in the motor cortex. These "upper motor neurons" are constantly sending signals that stimulate the "lower motor neurons" in the spinal cord, which causes the contraction of muscles. Neurons in the striatum function to inhibit signals from upper motor neurons that would otherwise result in unwanted body movements. The inhibitory neurons in the striatum, called "medium spiny neurons," use the neurotransmitter GABA, but in Huntington's disease the medium spiny neurons degenerate, resulting in uncontrollable body movements.

The question then became: Why were the GABAergic medium spiny neurons in the striatum selectively destroyed in the people who consumed the moldy sugar cane? Neuroscientists found that following exposure to 3-NPA, rats exhibit the uncontrolled body movements associated with the death of medium spiny neurons in the striatum (Ludolph et al. 1991). Further research showed that 3-NPA inhibits an enzyme in the mitochondria that is necessary for the production of ATP. Through this inhibition, 3-NPA causes an energy deficit in the neurons. Medium spiny neurons may be particularly vulnerable to 3-NPA because they have high firing rates and strong input from glutamatergic neurons and so require unusually large amounts of ATP to support their activity. Neuroscientists showed that drugs that block glutamate receptors can prevent the degeneration of medium spiny neurons in rats exposed to 3-NPA. Therefore, it was made clear that 3-NPA kills the neuron by an excitotoxic mechanism.

One of the most interesting stories in the history of neurology began in 1982. While working at the Santa Clara Valley Medical Center in San Jose California, the neurologist J. William Langston received a call from his chief resident, who told him that he was seeing an unusual patient. According to Langston,

> The patient's condition was indeed extraordinary. He was clearly awake, but had virtually no spontaneous movement, and exhibited "waxy flexibility" (when his arm was involuntarily raised it would stay in the position for a prolonged period of time). The answer came quickly. In my experience, when passively flexing the wrist or elbow of a catatonic patient, there is a distinct feeling of irregular active resistance. This patient had the "lead pipe rigidity" with the "cog-wheeling" of Parkinson's disease. Indeed he looked like a textbook case of advance PD before the days of levodopa. But this case didn't fit either answer. He was in his early forties, and his symptoms came on literally overnight. We had a first class medical mystery on our hands. (Langston 2017, S12)

During the next few days, six other young patients with identical symptoms came to area hospitals. The patients had a dramatic improvement in their symptoms when they were treated with levodopa, a drug that increases dopamine levels and is widely prescribed for Parkinson's patients. The question then became: What did these seven people have in common? They did not know each other, but they all were heroin users who had recently obtained a specific batch of synthetic designer heroin. Langston was able to get various samples of synthetic heroin confiscated by the police during raids, and analysis of the samples showed that although most were indeed heroin, the batch used by the "Parkinsonian" patients consisted almost entirely of the chemical 1-methyl-4-phenyl-1,2,3,6-tetrahydropyridine, or MPTP.

The next step was to see whether administration of MPTP to laboratory animals would cause symptoms similar to Parkinson's (Langston 2017). Within one day of administration of a single dose of MPTP, rats and mice developed a severe impairment in their ability to control their body movements. Examination of their brains revealed extensive destruction of dopaminergic neurons in the substantia nigra but no damage to other types of neurons. Further investigations revealed the surprising explanations for why and how dopaminergic neurons were selectively killed by MPTP. When MPTP enters the brain, it is taken up by astrocytes. The astrocytes have an enzyme called "monoamine oxidase B" that normally functions to degrade dopamine. This enzyme oxidizes (removes electrons from) MPTP to form MPP^+. MPP^+ is then extruded from the astrocytes and selectively transported into dopaminergic neurons but not into other neurons. The reason for this selectivity is that the axons of the dopaminergic neurons have a dopamine transporter in their membrane whose function is to take dopamine back into the axon after it has been released at the synapse. The transporter moves MPP^+ into dopaminergic neurons, where it accumulates at high concentrations.

MPTP may damage and kill dopaminergic neurons by causing alterations in their energy metabolism and impairing the function of mitochondria. As with all neurons in the brain, dopaminergic neurons are activated by glutamate. When dopaminergic neurons' energy levels are depleted by MPP^+, their ability to extrude Na^+ and Ca^{2+} is impaired. Because the dopaminergic neurons have high levels of glutamate receptors, when their mitochondria are disabled by MPP^+, they undergo excitotoxic damage and death.

Epidemiological studies have shown that occupational exposure to certain pesticides and herbicides increases the risk of Parkinson's disease. One pesticide that may contribute to some cases of Parkinson's disease is rotenone (Cicchetti, Drouin-Ouellet, and Gross 2009). This pesticide was isolated from the roots of the plant *Derris eliptica*, which grows in Southeast Asia. Many people still purchase rotenone and apply it to plants in their garden. Although it works well in preventing bugs from chewing on vegetables, when you eat those vegetables, you also ingest rotenone. Like MPP^+, rotenone reduces ATP production in cells by impairing electron transport in the mitochondria. High doses of rotenone are not acutely toxic because they cause vomiting and thus do not accumulate in the blood. However, repeated oral administration of moderate doses of rotenone to rats causes degeneration of dopaminergic neurons in their brains and associated difficulties in the rats' ability to control their body movements.

At first, it was thought that enough rotenone reaches the brain to directly cause degeneration of dopaminergic neurons. But further investigations have pointed to a surprising explanation: that rotenone first affects the neurons that innervate the intestines and control movement of food through the intestines. This is interesting because, as you will learn in chapter 8, recent evidence indicates that Parkinson's disease begins in the gut and makes its way to the brain through the axons of brainstem cholinergic neurons that innervate the gut.

The man-made chemical paraquat is among the most widely used herbicides throughout the world. Chemists first synthesized paraquat in 1882, but its ability to kill weeds was not established until the mid-1950s. High doses of paraquat are lethal, and there are many documented cases of suicide and a few cases of homicide resulting from paraquat ingestion. Agricultural workers exposed to paraquat have been reported to have a higher incidence of Parkinson's disease compared to people with nonagricultural occupations. Like MPP^+ and rotenone, paraquat blocks the electron transport chain in the mitochondria resulting in depletion of ATP and excitotoxic cell death. The chemical structure of paraquat is very similar to that of MPP^+. However, in contrast to MPP^+, the movement of paraquat into neurons does not require the dopamine transporter. Also in contrast to MPP^+, which almost exclusively kills dopaminergic neurons, paraquat kills other types of neurons as well.

THE DENDRITE'S PREDICAMENT

As detailed in chapter 3, the growth of dendrites is exquisitely sensitive to glutamate. Low concentrations of glutamate cause the dendrites to grow more slowly, and somewhat higher concentrations of glutamate cause the dendrites to regress. Still higher concentrations of glutamate cause the dendrites to degenerate—they become fragmented and disintegrate. Examinations of neurons in the brains of animals exposed to excitotoxins and of patients with epilepsy have shown that the excitotoxins destroy the dendrites of neurons but often spare the axons. Why are dendrites in such a precarious situation? The short answer is that the receptors for glutamate are located in the membrane of dendrites, where the receptors are concentrated at synapses.

At the synapse, the dendrite membrane bulges out from the shaft of the dendrite to form a dendritic spine (figures 1.1 and 1.3). Dendritic spines are most commonly shaped like a mushroom and have several features that make them hot spots for excitation by glutamate. Just inside the membrane of a dendritic spine is a scaffold made from several different proteins. The protein scaffold maintains the shape of the spine and holds glutamate receptors in place in the membrane. One of the most abundant scaffold proteins in the dendritic spine is postsynaptic density 95 (PSD95). Polymers of PSD95 form a sort of chain that runs parallel to the membrane surface and keeps glutamate receptors from moving laterally in the dendrite's membrane. Other proteins in the scaffold are involved in the insertion of glutamate receptors into the membrane or in the removal of glutamate receptors from the membrane. In addition, several enzymes are associated with the postsynaptic density protein scaffold and are involved in the transfer of information from the glutamate receptors to other proteins in the dendrite or to the cell nucleus.

Activation of glutamate receptors on one dendritic spine normally results in a large increase in Ca^{2+} levels in that spine, but with little or no increase in Ca^{2+} levels in adjacent spines on the same dendrite. The resting Ca^{2+} concentration inside a dendritic spine in which glutamate receptors are not activated is approximately 100 nanomolar, which is about 10,000 times lower than the extracellular Ca^{2+} concentration. When glutamate is released from an axon terminal onto a dendritic spine, the Ca^{2+} concentration in the spine typically increases approximately 100 times to about 10 micromolar. That high concentration of Ca^{2+} usually lasts for only a few seconds, and then the resting concentration returns. There are several mechanisms for the rapid removal of Ca^{2+} from the dendritic spine. The opening of K^+ channels results in movement of K^+ from the inside of the dendrite to the outside, which repolarizes the membrane. Also contributing to membrane repolarization is the activity of the Na^+ pump protein, which moves Na^+ out of the dendrite. This membrane repolarization results in the closing of NMDA receptor channels, thereby precluding further Ca^{2+} influx. Mitochondria may also contribute to the removal of Ca^{2+} from dendrites following the activation of glutamatergic synapses. In addition, dendrites contain calcium-binding proteins that can sequester Ca^{2+}.

It is thought that at least some calcium-binding proteins can protect dendrites against excitotoxic damage. Perhaps the strongest evidence for this comes from studies of the hippocampus. When rats or mice are exposed to kainic acid or domoic acid, the dendrites of CA3 pyramidal neurons and to a lesser extent CA1 pyramidal neurons degenerate, whereas the dendrites of dentate granule neurons do not. The dentate granule neurons' resistance to excitotoxicity is not due to the amounts or type of glutamate receptors they have. Indeed, dentate granule neurons have relatively high amounts of kainic acid and NMDA receptors. Their resistance to excitotoxicity is apparently not the result of their particular location in the hippocampal circuitry, either. Indeed, I performed a study that showed that intrinsic differences between dentate granule neurons and pyramidal neurons account for their differential vulnerability to excitotoxicity. The experiments involved removing the hippocampus from newborn rats, separating the different regions (CA1, CA2, CA3, and dentate gyrus) by microdissection, and culturing neurons from these different regions in separate culture dishes. The published article summarized the findings: "Experiments with specific glutamate receptor agonists and antagonists demonstrated that both non N-methyl-D-aspartic acid (NMDA) receptors and NMDA receptors mediated glutamate-induced degeneration. There were clear differences in the vulnerability of the neuron populations in cultures from the different hippocampal regions. The rank order of the vulnerability of pyramidal-like neurons to glutamate-induced

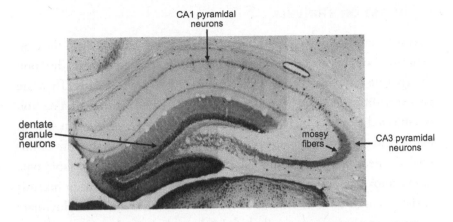

CA1 pyramidal
neurons

dentate
granule
neurons

mossy
fibers

CA3 pyramidal
neurons

Figure 6.1

The calcium-binding protein calbindin is produced in high amounts in hippocampal dentate granule neurons, in much lower amounts in CA1 pyramidal neurons, and in negligible amounts in CA3 pyramidal neurons. A slice of hippocampus from a rat was stained using an antibody to calbindin. The cell bodies, dendrites, and axons (mossy fibers) of granule neurons stain intensely, those of CA1 neurons lightly, and those of CA3 neuron not at all. Adapted from figure 2A in Sloviter (1989).

neurodegeneration between regions in culture was: DG [dentate gyrus] less than CA2 less than CA3 less than CA1" (Mattson and Kater 1989, 110).

One possible explanation for the remarkable resistance of dentate granule neurons to excitotoxicity is that they have high amounts of the calcium-binding protein calbindin (figure 6.1). Studies of cultured hippocampal neurons have shown that neurons with high amounts of calbindin are relatively resistant to dendritic damage and death compared to neurons with little or no calbindin (Mattson, Rychlik, et al. 1991). Dentate granule neurons with calbindin are also resistant to degeneration in patients with epilepsy and Alzheimer's disease. But calbindin is certainly not the only determinant of neuronal resistance to excitotoxicity because even dentate granule neurons in mice lacking calbindin remain relatively resistant to excitotoxicity. Other factors proposed to explain different susceptibilities of granule and pyramidal neurons to excitotoxicity include differences in levels of protein chaperones, such as heat-shock protein 70 and glucose-regulated protein 78; certain mitochondrial proteins; and levels of neurotrophic factors such as BDNF and FGF2.

Excessive activation of glutamate receptors can kill neurons. Both the concentration of glutamate to which the neuron is exposed and the duration of exposure influence the neuron's vulnerability to excitotoxicity. There are two very different types of such excitotoxic death of neurons. If a neuron is subjected to a very high amount of glutamate or an excitotoxin such as kainic acid or domoic acid for a sustained period exceeding about five minutes, the neuron will swell and burst within an hour or so. This rapid type of excitotoxic neuron death is called "necrosis." What happens in necrosis is that a continuous activation of the AMPA and kainic acid glutamate receptors results in massive influx of Na^+ through those glutamate receptors and voltage-dependent Na^+ channels, which results in a large increase in the concentration of Na^+ inside the neuron. Water then moves into the cell by osmosis, a process in which molecules of a solvent (water in this instance) pass through a semipermeable membrane from a less concentrated solution to a more concentrated solution. As a result, the cell swells, the membrane stretches to the point of rupture, and the cell contents are spilled.

When a neuron dies by necrosis and "spills its guts," the adjacent neurons are exposed to the dead neuron's contents. Because glutamate is normally concentrated inside of neurons, when a neuron dies by necrosis, the glutamate in that neuron is released and can activate glutamate receptors on adjacent neurons. In addition, enzymes released from the dying neuron can chew up proteins on the surface of adjacent neurons, which compromises these neurons' ability to maintain the electrical potential across the membrane. Another major adverse effect of necrosis on surrounding cells is inflammation. Normally, the immune cells—microglia—present in the brain are not exposed to proteins, DNA, and other molecules that are inside neurons. When these molecules are released from a dying neuron, however, the brain's immune cells respond in the way they are supposed to respond and so gobble up the remnants of the necrotic neurons, but in doing so the microglia release several substances that render living neurons vulnerable to excitotoxicity.

Three substances released by immune cells that contribute to excitotoxicity are superoxide, hydrogen peroxide, and nitric oxide. All three of these substances cause oxidative damage to neurons. Superoxide and nitric oxide are free radicals, and hydrogen peroxide interacts with iron ions (Fe^{2+}), resulting in the production of hydroxyl radical. Hydroxyl radical and nitric oxide are particularly damaging to cells because in a process called "lipid peroxidation" they attack the fats that form the membrane of cells. During this process, a small piece of membrane lipid called "4-hydroxynonenal" (HNE) is released and binds to proteins in the cell membrane and thereby impairs their function. Several of the proteins to which HNE binds are important for their ability to prevent excitotoxicity, including those proteins that move Na^{2+} and Ca^{2+} out of the neuron and a protein that moves glucose into the neuron (Mark, Lovell, et al. 1997; Mark, Pang, et al. 1997).

Chapter 3 described how during brain development many neurons die by the natural process of apoptosis. In contrast to necrosis, where the death of one neuron can cause neighboring neurons to die, when a neuron dies by apoptosis, it does not adversely affect its neighbors. Moderate sustained levels of glutamate receptor activation can cause a neuron to die by apoptosis. Glutamate causes a sustained elevation of intracellular Ca^{2+} levels, which results in stress on mitochondria. The mitochondria then release the protein cytochrome C, which in turn causes the activation of enzymes called "caspases." The caspases degrade cellular proteins in a coordinated manner that causes the cell to shrink and be gobbled up by microglia without the neuron ever having spilled its guts. This is why when a neuron dies by apoptosis, it does not adversely affect neighboring neurons.

Neurons are particularly vulnerable to apoptosis caused by glutamate receptor overactivation when they experience a deficit in neurotrophic factors. Studies have shown that BDNF, FGF2, and IGF1 can keep neurons alive when they are exposed to amounts of glutamate that would otherwise cause apoptosis (Mattson, Murrain, et al. 1989; B. Cheng and Mattson 1992b, 1994). These neurotrophic factors affect the expression of glutamate receptors and calcium-binding proteins, bolster mitochondrial stress resistance, and stimulate the production of antioxidant enzymes. There is evidence that neurotrophic factor levels decrease in the brain during aging, which would be expected to increase the neurons' vulnerability to apoptosis triggered by glutamate receptor overactivation.

Both excitotoxic necrosis and apoptosis are thought to occur in epilepsy, stroke, and traumatic brain injury. Neurons that are most severely affected by these conditions undergo rapid necrosis, while those that are less severely affected undergo delayed apoptosis. Apoptosis may occur in slowly progressing neurodegenerative disorders such as Alzheimer's and Parkinson's diseases. Chapters 7 and 8 describe roles for neuronal network hyperexcitability and excitotoxicity in these brain disorders.

7 UNEXPECTED ATTACKS

Neurons in the brain can experience sudden damage and death in three common circumstances—epileptic seizures, stroke, and head trauma. Depending on the brain regions affected and the extent of damage, the outcome of these unexpected insults ranges from little or no discernable functional consequences to severe long-term disability to sudden death. The vast majority of neurons affected in epilepsy, stroke, and traumatic brain injury are glutamatergic. Epileptic seizures are usually unpredictable and can affect anyone regardless of their age and general health. In contrast, there are clear and common risk factors for stroke and a traumatic brain injury. To reduce your risk of a stroke, you can maintain a low body weight, exercise regularly, eat a healthy diet, and control your blood pressure. To avoid traumatic brain injury, you can wear a seat belt when driving a car and a helmet when riding a bicycle or motorcycle. Parents should consider not having their child participate in contact sports such as football, where head trauma is common.

EPILEPSY

The recorded history of epilepsy begins more than 2,500 years ago in what is now Iraq. Someone there carved a clay tablet that shows a demon with a long tail, curved horns, and a serpent's tongue. Above the demon is cuneiform writing that describes a condition the Assyrians called *bennu*, in which the affected person exhibits convulsions. Another very old description of a person with what was presumably epilepsy comes from ancient Mesopotamia and dates to approximately 2000 BCE. The affected person was thought to be in the grip of a "moon god" and so was subjected to an exorcism. A Babylonian medical text from about 1050 BCE describes signs and symptoms consistent with epilepsy and considered the seizures to be the result of possession by demons. The ancient Greeks, however, believed that people with epilepsy were highly intelligent and had divine powers. But Hippocrates, who is now considered the father of medicine, believed that epilepsy resulted from a problem in the brain and described several children with severe recurrent seizures. In ancient Rome, it was common for physicians to try to elicit an epileptic seizure by having the patient watch a spinning potter's wheel, which is consistent with the fact that in some instances a seizure can be induced by flashing light.

Many people are affected by epilepsy. During the last year of his life and while still in office, President Franklin Roosevelt of the United States developed epileptic seizures, which were kept secret from the public. Since then, several people have described his behavior during the seizures. For example, the reporter Turner Catledge wrote:

> When I entered the president's office . . . he was sitting there with a vague glassy-eyed expression on his face and his mouth hanging open. He would start talking about something, then in midsentence he would stop and his mouth would drop open and he'd sit staring at me in silence. . . . Repeatedly he would lose his train of thought, stop, and stare blankly at me. It was an agonizing experience for me. Finally a waiter brought his lunch, and [Chief of Staff, General Edwin "Pa"] Watson said his luncheon guest was waiting, and I was able to make my escape. (1971, 146)

The cause of Roosevelt's seizures was thought to be related to cerebral vascular abnormalities as he eventually died from a stroke.

The entertainer Prince had childhood epilepsy, which resolved and did not recur during his adult life. This is fairly common. Many children who have seizures will become seizure free as they grow up. The Australian actor Hugo Weaving, who starred in *The Matrix*, *Lord of the Rings*, and *V for Vendetta*, developed epilepsy when he was 13 and had seizures every year until he was in his early forties. Before Weaving developed epilepsy, he was very shy and had much anxiety when around people other than his immediate family. When he was put on a high dose of drugs to treat his epilepsy, his anxiety was greatly reduced, and he was able to realize his dream of becoming an actor.

Florence Griffith-Joyner—"Flo Jo"—was a sprinter who won a silver medal for the United States in the 200 meters in the 1984 Olympics and gold medals in the 100 and 200 meters in the 1988 Olympics, setting a world-record time in the 200 meters. On September 21, 1998, Flo Jo died in her sleep at the age of 38. The coroner concluded that she died of suffocation as the result of a severe epileptic seizure.

Approximately 3.5 million people in the United States have epilepsy, of which about 500,000 are children. A person with epilepsy experiences recurrent seizures that result from unconstrained vigorous activity in neuronal networks in one or both hemispheres of their brain. The seizures most commonly occur spontaneously and without any obvious cause. In some instances, seizures occur after a stroke or traumatic brain injury or as a consequence of a brain tumor. Seizures can occur at any age but are most common in children and older people. Neurologists have classified seizures into different types based on the symptoms. In the most common type of seizure, the person will manifest convulsions. In some cases, the person will experience an "absence seizure," during which they have an altered consciousness. Some seizures are preceded by an "aura," in which the person experiences a sensation such as a smell, sound, or visual image. Seizures generally last for several seconds to a minute or more, during which time the person is vulnerable to injury or even death. For example, a seizure that strikes while the person is driving a car may cause a crash.

Overactivation of glutamate receptors in brain neuronal networks is both necessary and sufficient to cause epileptic seizures. Drugs that block glutamate receptors can prevent seizures in animal models of epilepsy. However, the drugs most commonly prescribed to treat epilepsy are not glutamate receptor antagonists because blocking glutamate receptors can cause side effects such as impaired learning and memory. Drugs prescribed for epilepsy instead reduce the excitability of neurons by either blocking voltage-dependent Na^+ channels or activating inhibitory GABA receptors or inhibiting voltage-dependent Ca^{2+} channels. Anticonvulsant drugs that reduce excitability by inhibiting Na^+ channels include phenytoin, valproic acid, and carbamazepine. Benzodiazepines such as diazepam and clonazepam activate GABA receptors, and gabapentin inhibits Ca^{2+} channels.

Although the vast majority of cases of epilepsy are not inherited and the cause is unknown, there are rare cases caused by a genetic mutation. Not surprisingly, the affected genes encode proteins involved in the regulation of neuronal excitability (Szepetowski 2018). Several different mutations in K^+ and Na^+ channels can cause inherited epilepsy. Mutations in GABA receptors are responsible for other instances of inherited epilepsy. Children with a mutation in a subunit of the NMDA type of glutamate receptor exhibit seizures at an early age and may also have intellectual disabilities. Interestingly, children with autism are more likely to have seizures compared to those without autism. Chapter 9 describes emerging evidence suggesting that neuronal network hyperexcitability occurs in autism and results from abnormalities that arise during brain development in the mother's womb.

Beyond alterations in the function of ion channels, research suggests important roles for mitochondria in epilepsy. Evidence indicates that robust mitochondria can protect neurons against seizures. There are hundreds of different proteins in mitochondria, but recent discoveries suggest a particularly important role for the protein SIRT3 in protecting neurons against hyperexcitability and seizures.

SIRT3 is an enzyme that removes an acetyl group from the amino acid lysine in proteins in a process called "deacetylation." Enzymes known as acetyltransferases can add an acetyl group to lysines in proteins. Many proteins have one or more lysines, and these proteins' function is often affected by the presence or absence of an acetyl group. Thus, protein acetylation is analogous to the more widely studied process of protein phosphorylation, in which kinases add and phosphatases remove a phosphate group from certain amino acids of proteins. Research findings suggest that more than 100 proteins in mitochondria are deacetylated by SIRT3. Some of these proteins are part of the electron transport chain involved in ATP production, whereas others are antioxidant enzymes that remove the free radicals produced during electron transport. In general, SIRT3 improves the efficiency of mitochondria and protects them against stress. In these ways, SIRT3 may support the formation, maintenance, and proper function of synapses in the developing and adult brain.

In one study, Aiwu Cheng found that neurons in the brains of mice genetically engineered to lack the mitochondrial protein SIRT3 are highly vulnerable to seizures (A. Cheng, Yang, et al. 2016). More neurons in the hippocampus of SIRT3-deficient mice were killed by the seizure-inducing excitotoxin kainic acid compared to mice with normal levels of SIRT3. A mitochondrial energy deficit can increase the vulnerability of neurons to excitotoxicity. Several of the proteins in the mitochondrial electron transport chain are known to be deacetylated by SIRT3. Cheng found that ATP levels in the cerebral cortex and hippocampus were significantly lower in SIRT3-deficient mice. Therefore, compromise of the proteins' function likely explains the reduction in ATP levels in the brains of mice that lack SIRT3. Two additional proteins were identified as likely contributing to the susceptibility of SIRT3-deficient neurons to excessive accumulation of free radicals and Ca^{2+}. One protein is the mitochondria antioxidant enzyme superoxide dismutase 2 (SOD2), which removes superoxide free radicals. SIRT3 increases the activity of SOD2, thereby enhancing removal of superoxide free radicals. The other protein is cyclophilin D, which is involved in the formation of pores in the mitochondrial membrane and the triggering of apoptosis. Deacetylation of cyclophilin D by SIRT3 prevents formation of the pores and consequent accumulation of Ca^{2+} in the mitochondria. By keeping free radical levels low and stabilizing the mitochondrial membrane, SIRT3 protects neurons against excessive Ca^{2+} accumulation when they are stimulated by glutamate.

Seizure incidence is higher in older people compared to middle-aged people. Several features of the aging brain may enable the development of seizures (Mattson and Arumugam 2018). During aging, neurons experience increased oxidative damage to DNA and proteins as a result of dysfunction of mitochondria, a decline in antioxidant defenses, a reduced ability to repair the damage, and an impaired ability to remove damaged proteins and mitochondria (figure 7.1). The proteins damaged by oxygen free radicals include those that prevent neuronal hyperexcitability. Age-related oxidative damage to the Na^+ and Ca^{2+} pump proteins in the neurons' membrane impairs the

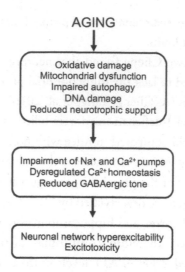

AGING

Oxidative damage
Mitochondrial dysfunction
Impaired autophagy
DNA damage
Reduced neurotrophic support

Impairment of Na⁺ and Ca²⁺ pumps
Dysregulated Ca²⁺ homeostasis
Reduced GABAergic tone

Neuronal network hyperexcitability
Excitotoxicity

Figure 7.1

Some of the changes that occur in neurons during aging that render them vulnerable to hyperexcitability and excitotoxicity.

neurons' ability to remove those ions after the membrane depolarization caused by glutamate receptor activation. Another protein damaged by oxidative stress is a glutamate transporter protein in astrocytes. The latter protein normally functions to remove glutamate from a synapse after a neuron has fired an action potential. Its impairment therefore increases the amount of glutamate at the synapse and so makes neuronal networks prone to excessive excitation and seizures. Neurons' ability to generate ATP is diminished during aging as a result of damage to mitochondria and an impaired ability to utilize glucose as an energy source. In addition, there is evidence that neurotrophic factor support declines during aging. Collectively, these adverse effects of aging on brain cells promote neuronal network hyperexcitability and may trigger seizures.

STROKE

The actress Grace Kelly played leading roles in many widely acclaimed movies in the 1950s, including *Rear Window, Dial M for Murder, The Country*

Girl, and *Mogambo.* On September 13, 1982, Kelly was driving with her daughter in Monaco when her car unexpectedly veered off the road and over a cliff. The next day she died. But the extent of the injuries she suffered in the crash could not explain her death. Her daughter recalled that her mother had complained of a headache and then suffered sudden pain before the car went over the cliff. An autopsy revealed that Grace had suffered two strokes, one that preceded and caused the crash and another after the crash.

As Charles Dickens wrote *Great Expectations, A Tale of Two Cities, A Christmas Carol,* and *Oliver Twist,* insidious events were occurring in the arteries in his brain. On April 18, 1869, at the age of 57 Dickens suffered a stroke and collapsed four days later. He recovered sufficiently to work on his final novel, *The Mystery of Edwin Drood.* In early 1870, Dickens resumed public performances in a series of readings at different locations in London and surrounding areas. On the evening of June 8, 1870, he suffered a massive stroke and died the next day.

On June 23, 1953, Prime Minister Winston Churchill of Great Britain suffered a stroke that resulted in partial paralysis of one side of his body. He nevertheless held a cabinet meeting the next morning without anyone noticing his disability. In the ensuing days, Churchill's condition worsened, and he went to his home in Chartwell. His condition was kept secret from Parliament and the public, and he recovered during a four-month period. Churchill continued on as prime minister until April 1955, when he retired. Ten years later he suffered another stroke and died two weeks later at the age of 87. Several factors in Churchill's life likely contributed to his strokes, including that he was overweight, smoked, drank, and experienced much stress as prime minister.

Stroke is the fifth leading cause of death in the United States and a major cause of long-term disability. Each year approximately 800,000 people in the United States suffer a stroke, which means that several people are having a stroke while I write this paragraph. It is estimated that the yearly cost of stroke is $50 billion. Most strokes result from a clot forming in an artery in the brain, referred to as an "ischemic stroke." About 10 percent of strokes result from the rupture of an artery, called a "hemorrhagic stroke." The symptoms of stroke are directly related to which artery is affected and the extent of damage to neurons in the brain region perfused by that artery. The most commonly affected artery is the middle cerebral artery, which supplies blood to the striatum and areas of the cerebral cortex that include the motor cortex and sensory cortex. The symptoms reveal which side of the brain was affected by the stroke because the left side of the brain controls the right side of the body, and the right side of the brain controls the left side of the body. If the stroke occurs on the left side of the brain, the damage to neurons in the motor cortex and striatum will cause problems in controlling the movement of muscles on the right side of the face and body. When the sensory cortex on the right side of the brain is affected, the person will have numbness in the contralateral face and/or limbs.

The changes in a cerebral artery that predispose it to occlusion by a clot include inflammation, the accumulation of macrophages, and the deposition of cholesterol in an ever-growing plaque that narrows the artery. Platelets are attracted to the site of inflammation and can form a clot that results in a dramatic decrease or complete blockage of blood flow through the artery. The medical term for such a severe decrease in blood flow is "ischemia."

Because neurons are electrically active and have a high demand for energy (glucose and ketones) and oxygen, they are in a precarious situation when their blood supply is interrupted. In fact, even a minute or two of ischemia can kill a neuron. Most commonly, a clot that forms in an artery will resolve within many minutes to a few hours after it has formed, and blood flow will then resume, a process called "reperfusion." Evidence suggests, though, that many neurons die after reperfusion when the resupply of oxygen results in a profound increase in free radical levels.

The amount of brain tissue that receives nutrients and oxygen from an artery is that artery's "perfusion territory." There is often considerable overlap in the territories of different arteries, which can enable neurons in those "watershed areas" to survive a stroke. Examinations of the brains of people who died from an ischemic stroke and studies of animal models of stroke have revealed two areas of damage distinguished by the type and extent of neuronal death. In a "core" area that receives its blood supply exclusively from the occluded artery, all of the neurons die rapidly by necrosis. In a surrounding area called the "penumbra," neurons die more slowly by apoptosis. Studies of animal models of stroke have shown that the delayed death of neurons in the penumbra can be prevented by certain interventions, including drugs that block glutamate receptors.

To simulate a stroke in a rat or mouse, a surgical procedure is used to temporarily occlude the middle cerebral artery. The extent of the brain damage and the severity of symptoms that result from the experimental stroke depend on how long the blood flow is halted. For example, blocking the middle cerebral artery for one hour causes extensive death of neurons in the striatum and somewhat less damage in the penumbra region in the cerebral cortex. Damage to neurons in the striatum and motor cortex results in a partial paralysis in the opposite side of the body and a prominent impairment in the animal's ability to walk normally. Studies in the late 1980s using this animal model showed that the death of many neurons in the penumbra area can be prevented by treating the animal with NMDA receptor antagonists such as dizocilpine, ketamine, and dextromethorphan.

Unfortunately, clinical trials of drugs that block glutamate receptors in human stroke patients have thus far failed to show a benefit. In fact, except for the clot-busting drug tissue plasminogen activator (tPA), which is beneficial in only about 5 percent of cases, all drug trials in stroke patients have failed. There are several likely reasons why a drug is found to be beneficial in an animal model of stroke but then fails to show a benefit in humans. First, there is tremendous variability among human stroke patients with respect to which artery in their brain is affected, how long it is occluded before the clot resolves, and their general health. In contrast, in the animal models the blood flow is blocked in the exact same part of the middle cerebral artery for the exact same time for all animals. Second, in contrast to humans, laboratory rats and mice are inbred with little or no genetic variability between individuals, and they all are raised in the same environment. This means that there is much less interindividual variation in the amount of brain damage caused by experimental stroke in animals compared to stroke in humans. Third, in animal studies the drug being tested is often administered prior to or shortly after the stroke, whereas in human studies the timing of drug administration depends on how soon the person gets to the hospital and is diagnosed and then treated with the drug. Fourth, in animals the arteries supplying blood to the brain are undamaged in contrast with the arteries in humans, which in most instances are damaged by hypertension, atherosclerosis, and inflammation.

But neuroscientists' efforts to understand how activation of glutamate receptors contributes to the death of neurons in stroke are revealing potential approaches to keeping the neurons alive despite hyperactivation of glutamate receptors. Key to these efforts are studies that elucidate how the Ca^{2+} influx resulting from excessive glutamate receptor activation triggers the biochemical events that result in the destruction of the neuron. Evidence suggests that the excessive Ca^{2+} influx adversely affects mitochondria, resulting in apoptosis. Thus, bolstering mitochondrial stress resistance might prove beneficial in treating stroke. For example, the drug cyclosporine A blocks the membrane permeability transition pores, whose opening triggers the activation of caspases and apoptosis. Cyclosporine A rescues neurons in the ischemic penumbra and ameliorates behavioral deficits in animal models of stroke (Forsse et al. 2019). A recent clinical trial showed that cyclosporine A reduces the amount of brain damage in some stroke patients (Nighoghossian et al. 2015). Methylene blue is a chemical that can simultaneously increase mitochondrial ATP production and mitigate oxidative stress. It has proven effective in reducing neuronal degeneration and improving functional outcomes in animal models of stroke (Tucker, Liu, and Zhang 2018).

Another approach to keeping neurons alive after a stroke is to provide them with ketones. The ketones beta-hydroxybutyrate and acetoacetate are produced from fats during fasting or from ketogenic foods. Several studies have shown that intermittent fasting prior to a stroke can reduce brain damage and improve functional outcome in rats and mice (Yu and Mattson 1999; Arumugam et al. 2010). Moreover, many studies have shown that ketones can protect neuronal networks against epileptic seizures and consequent excitotoxicity. It turns out that ketones can also reduce the production of free radicals that occurs in neurons after a stroke because less free radicals are produced during the metabolism of ketones compared to the robust production of free radicals during the metabolism of glucose. Canadian researchers showed that administration of the ketone beta-hydroxybutyrate can reduce free radical levels in the brain tissue affected by an experimental stroke in rats (Bazzigaluppi et al. 2018). The rats given the ketone exhibited better functional recovery, suggesting a potential for ketone-based therapies in stroke patients.

Finally, evidence suggests that reducing the activity of the adrenal stress hormone cortisol may be beneficial in stroke patients. Robert Sapolsky and colleagues at Stanford University showed that chronic elevation of corticosterone (the rodent equivalent of cortisol) levels renders neurons vulnerable to excitotoxic damage (Stein-Behrens et al. 1994). When Ginger Smith-Swintosky was a postdoc, she and others in my laboratory found that a drug that prevents cells in the adrenal gland from producing corticosterone can protect neurons from dying in a rat model of ischemic stroke (Smith-Swintosky et al. 1996). This was an important finding because stroke patients had previously been treated with cortisol-like steroids, such as dexamethasone, with the idea that these steroids would reduce inflammation in the brain, when in fact they were worsening the damage to neurons caused by the stroke. The prescription of corticosteroids for stroke patients is now contraindicated.

BRAIN AND SPINAL CORD INJURIES

When my family and I lived in Lexington, Kentucky, one of my son's friends in elementary school was a boy named Howard. Shortly after we moved to the Baltimore area in 2000, Howard moved with his family to the Washington, DC, area. One day the next year Howard's parents contacted us to inform us that Howard had been hit by a car speeding through their neighborhood and was in critical condition in a coma due to severe head injury. Howard was transferred to the Johns Hopkins Kennedy-Krieger Institute in Baltimore, where we visited him every weekend for two months. After about two weeks, he came out of the coma but was unable to talk and could barely move his limbs. During the ensuing month, he became able to talk with slurred speech and to stand and walk with assistance. He eventually regained speech, albeit with some slurring, and was able to walk and perform the activities of daily living with some limitations. Howard graduated from high school and eventually earned a college degree.

Each year, nearly 3 million people in the United States will receive medical treatment for a traumatic brain injury (TBI), with more than 50,000 of them dying as a result. The most common cause of TBI is a fall, and the second most common cause is a vehicle crash. The disability resulting from TBI is considerable and in many cases severe enough to prevent the person from continuing their usual activities, such as work. Approximately half of all people with a TBI will experience major depression. Children who suffer a TBI often have difficulty in school. They may not be able to concentrate or may be prone to emotional outbursts. However, a child's brain is much more resilient than an older person's brain, and many children make remarkable recoveries from severe brain traumas that would likely have killed an older person.

On February 25, 1965, Muhammad Ali knocked out Sonny Liston to become the heavyweight boxing champion of the world. After refusing to be drafted into the military for the Vietnam War, he was stripped of his title and banned from boxing for three years. On March 8, 1971, he lost a ferocious bout with Joe Frazer. In a fight in Zaire on October 30, 1974, he knocked out George Foreman to regain the world heavyweight title. An intelligent, witty, and charismatic person, Ali was an outspoken advocate for civil rights. At the age of 42, though, he was diagnosed with Parkinson's disease. Ali's Parkinson's was very likely the result of his suffering numerous blows to the head during his 61 professional fights, particularly later in his career. Toward the end of his life, Ali also suffered from impaired cognition.

Neurologists have used the term *dementia pugilistica* to describe the brain damage and resulting behavioral and cognitive symptoms in boxers. The condition, now called "chronic traumatic encephalopathy," results from multiple concussions and has recently been found to be common in US football players. Neuropathologists have examined the brains of deceased former professional football players who suffered concussions during their careers and found abnormalities consistent with degeneration of neurons in several regions, including the frontal cortex (McKee 2020). Many of these players suffered from depression and anger as well as cognitive impairment. When players became aware of this danger, National Football League management was forced to change some of the game's rules to reduce players' risk of suffering a concussion.

Some people who suffer a TBI are able to function at a high level but with clear changes in some cognitive or emotional domains. This was the case for the British writer Roald Dahl, who is best known for his works for children, including *Charlie and the Chocolate Factory*, *Matilda*, *The BFG*, and *James and the Giant Peach*. Dahl was a fighter pilot in World War II. When his plane crashed in 1940, he suffered multiple injuries, including a TBI. After recovering, he experienced a personality change that included increased confidence, a reduced sense of embarrassment, and pleasure in shocking people. This personality change translated to the considerable dark humor in his post-TBI writing that was not present in his previous works.

TBI occurs when a single hard blow or multiple less severe blows to the head causes physical damage to brain cells, which may be concentrated in one area of the brain or may be more diffuse. The sites of severe trauma can be seen on MRI images as "bright" areas, showing where fluid has accumulated. This accumulation is referred to colloquially as "brain swelling." Insight into the biochemical changes that occur in the brains of TBI patients has been gained by studying cerebrospinal fluid collected via spinal taps. This fluid circulates outside of cells throughout the brain. It contains chemicals released by brain cells. By analyzing the cerebrospinal fluid of TBI patients in samples taken at different points after the concussion occurs, neurologists have found that glutamate levels in the fluid are increased soon after a TBI and remain elevated for at least a week in many patients. Damage to the membrane of neurons causes glutamate to leak out, resulting in an increase in the extracellular glutamate concentration. In addition, the ability of astrocytes to remove glutamate is thought to be compromised in TBI.

Evidence from studies of rats and mice indicates that excitotoxicity resulting from excessive activation of glutamate receptors and consequent Ca^{2+} accumulation inside the neuron contributes to the brain damage and symptoms of TBI. When administered immediately after head trauma, drugs that block the NMDA receptor reduce the amount of brain damage in animal models of TBI (Smith et al. 1993).

Roy Campanella was born in Philadelphia in 1921. His mother was African American, and his father the son of Italian immigrants. Roy attended integrated schools and excelled in all sports, among which baseball was his favorite. At the age of 16, he had the opportunity to play professional baseball in the Negro League and so dropped out of high school to do so. Roy then moved to the Mexican League, where it became clear that his abilities were as good or better than most players in Major League Baseball. Finally, in 1947, when Jackie Robinson broke the color barrier, Campanella signed a contract with the Brooklyn Dodgers. his performance with the Dodgers established him as one of the best catchers in baseball history. In the winter of 1958, Campanella was driving from Harlem to Glen Cove, New York, when his car skidded on a patch of ice and struck a telephone poll, which resulted in the fracturing of his spine, leaving him paralyzed below his shoulders. He was confined to a wheelchair the rest of his life.

The DC Comics character Superman was faster than a speeding bullet, stronger than a locomotive, and able to fly. He was born on the planet Krypton and was sent to Earth by his father as Krypton was about to be destroyed in a war. On Earth, he caught criminals and prevented crimes. In 1978, the actor Christopher Reeve played the title role in the major box-office hit *Superman*. He also played the part of Count Aleksei Vronsky in the movie *Anna Karenina*, and as part of his role he learned to ride a horse. He continued to ride horses for pleasure after the movie and began competing in jumping events. On May 27, 1995, Reeve was thrown from his horse while competing in an event in Vermont, resulting in major damage to his upper spinal cord and paralysis of his entire body below the neck. Reeve was confined to a wheelchair and devoted the remaining years of his life to raising both awareness of spinal cord injury and money for research aimed at restoring function in people with this injury.

Approximately 250,000 people in the United States are currently living with a disability resulting from a spinal cord injury. Most of these injuries result from an automobile accident. Prognosis is generally poor because at the injury site the axons that traverse the spinal cord are damaged or severed. Those neurons' cell bodies are either in the brain (upper motor neurons), spinal cord (lower motor neurons), or dorsal root ganglia (sensory neurons).

Even if the neurons survive the trauma, their axons do not regenerate and so are disconnected from the peripheral organs they innervate. The axons of motor neurons are located in the ventral (front) half of the spinal cord, while those in sensory pathways are in the dorsal (back) half. Therefore, damage to the dorsal half of the cord results in sensory deficits; damage to the ventral half results in paralysis; and damage to both causes sensory and motor deficits.

At any level of the spinal cord, there are thousands of axons. As nerves branch off the spinal cord, the numbers of motor and sensory axons are greater at the upper (cervical and thoracic) regions than at the lower (lumber and sacral) regions. Accordingly, an injury to the upper spinal cord can cause complete paralysis of the body, while an injury to the sacral cord may cause only partial paralysis of the lower legs. Whereas neurons at the site of the injury may die rapidly, axons coursing through the spinal cord and the oligodendrocytes that myelinate them often undergo delayed degeneration during a period of several weeks after the injury.

As in traumatic brain injury and stroke, excitotoxic mechanisms are implicated in the demise of both neurons and oligodendrocytes in spinal cord injury. The tissue damage results in increased free radical production, mitochondrial dysfunction, and disruption of cellular ion homeostasis. Neurons become depolarized as a result of glutamate receptor activation and excessive Na^+ influx. As a consequence, voltage-dependent Ca^{2+} channels open, and the elevation of the Ca^{2+} concentration in the axon impairs mitochondrial function and activates Ca^{2+}-dependent proteases, which degrade cytoskeletal proteins, thereby contributing to the degeneration of the axon.

Whereas most AMPA receptors primarily flux Na^+, oligodendrocytes have a type AMPA receptor that is highly permeable to Ca^{2+}. Studies have shown that exposure of nerves with myelinated axons to glutamate can damage oligodendrocytes and impair the propagation of the action potential along the axons they ensheathe. This adverse effect of glutamate on oligodendrocytes can be prevented with drugs that block AMPA receptors. Studies have shown that drugs that block Na^+ channels or AMPA receptors can reduce damage to axons and lessen disability in animal models of spinal cord injury. Riluzole is a particularly promising drug that targets excitotoxicity. It has been shown to inhibit Na⁺ channels as well as enhance glutamate uptake by astrocytes. A preliminary clinical trial involving 36 people with spinal cord injury demonstrated a significantly greater improvement in motor function in those treated with riluzole compared to those treated with a placebo (Grossman et al. 2014). Although not yet approved for spinal cord injury, riluzole is widely prescribed to people with ALS.

Another approach for reducing excitotoxic degeneration of neurons in spinal cord injury is to administer neurotrophic factors into the area of injury by either infusion of cells that produce the neurotrophic factor or expression of the gene encoding the neurotrophic factor in glial cells. Preclinical studies in rat models of spinal cord injury have provided evidence that this is a promising therapeutic approach. For example, one study showed that continuous infusion of FGF2 into the spinal fluid with a minipump for one week after lumbar spinal cord injury resulted in a significant recovery of hindlimb movement during the subsequent five weeks (Rabchevsky et al. 2000). Another study reported that implantation of fibroblasts producing BDNF into the injured area improves functional recovery after cervical spinal cord injury in rats (Murray et al. 2002).

Inflammation occurs in stroke as well as in traumatic brain and spinal cord injuries. Microglia and infiltrating macrophages gobble up apoptotic cells and the cellular debris from cells that died by necrosis. However, the activated microglia and macrophages also produce superoxide and nitric oxide free radicals, which can damage otherwise healthy neurons.

Until the early 1990s, it was thought that the activation of immune cells in the brain was just detrimental to neurons. But then evidence emerged that tumor necrosis factor (TNF), a cytokine produced by microglia and macrophages, can protect neurons against the kinds of metabolic and excitotoxic stress that occur in stroke, epilepsy, and traumatic brain and spinal cord injuries.

Bin Cheng discovered that when he exposed cultured hippocampal neurons to TNF, their vulnerability to glutamate excitotoxicity was reduced. He also found that pretreatment with TNF protected hippocampal neurons from being damaged and killed by glucose deprivation (B. Cheng, Christakos, and Mattson 1994). But what is the mechanism by which TNF protects neurons against excitotoxicity? Steve Barger discovered that TNF activates the transcription factor NF-κB and provided evidence that activation of NF-κB explains TNF's ability to prevent neuronal death (Barger et al. 1995). Subsequent findings indicated that NF-κB increases the production of the mitochondrial antioxidant enzyme SOD2 in neurons, which likely contributes to TNF's "excitoprotective" effect (Bruce-Keller, Geddes, et al. 1999). Anna Bruce and colleagues found the neurons in the hippocampus of TNF-receptor-deficient mice were more vulnerable to being killed by kainic acid compared to neurons in control nondeficient mice. In the same study, Mark Kindy found that the amount of brain damage caused by an experimental stroke was greater in the mice lacking TNF receptors than in wild, nondeficient mice (Bruce et al. 1996). Altogether, the available evidence suggests that TNF plays an important role in protecting neurons against death caused by epileptic seizures and a stroke.

TNF levels increase dramatically in brain injuries, but normal levels of TNF and NF-κB activation may play important roles in the adaptive responses of glutamatergic neuronal networks to changes in neuronal activity. As evidence, Ben Albensi found that LTD at hippocampal CA1 synapses is impaired in mice lacking TNF receptors and provided evidence for a role for NF-κB in LTD (Albensi and Mattson 2000). Prolonged changes in a neuron's electrical activity can result in uniform changes in the strength of all synapses on that neuron. This process is called "homeostatic synaptic scaling" and is thought to enhance the performance of neuronal networks. Working at Stanford University, David Stellwagen and Robert Malenka showed that TNF mediates the synaptic scaling that occurs on glutamatergic hippocampal neurons in response to sustained inhibition of neuronal activity (Stellwagen and Malenka 2006).

In different studies, Michal Schwartz and Jonathan Kipnis have revealed roles for "T cells," a type of circulating immune cell, in the brain's adaptive responses to injury. In experimental animal models of multiple sclerosis and traumatic injury, when they depleted T cells from mice, damage to neurons was increased, but when they replenished the T cells, the damage was reduced (Schwartz and Raposo 2014). Remarkably, mice deficient in the T cells exhibit impaired learning and memory as well as altered responses to anxiety-provoking situations, which suggests important roles for these immune cells in normal brain functions, including cognition and adaptation to stress (Salvador, de Lima, and Kipnis 2021).

Activation of the immune system is a normal and important feature in the repair of injured tissue, and the results of the studies described in the preceding paragraphs suggest that at least some inflammation-related processes play important roles in protecting neurons against excitotoxicity. However, chronic inflammation may create an environment that renders neurons vulnerable to progressive dysfunction and degeneration. For example, it is thought that chronic "neuroinflammation" contributes to the demise of neurons in Alzheimer's disease. The next chapter considers the roles of glutamate in this and other insidious neurodegenerative disorders.

8 EVE OF DESTRUCTION

Alzheimer's disease and Parkinson's disease are the two most common neuro-degenerative brain disorders. Most cases of these degenerative brain diseases manifest in people who are in their seventies or eighties. Advances in the early diagnosis and treatment of cardiovascular disease, diabetes, and cancers have enabled many people who would have otherwise died from these diseases in their fifties and sixties to live for another decade or two and thus to enter the "danger zone" ages for Alzheimer's and Parkinson's diseases. These diseases not only are devastating for the affected person but also exact a heavy burden on caregivers because the affected person will become unable to perform the activities of daily living. Unfortunately, there are currently no treatments that can stop or even slow the progression of either disease. Both Alzheimer's and Parkinson's diseases involve an insidious type of "slow excitotoxicity" resulting from impaired energy metabolism in brain cells combined with the accumulation of neurotoxic proteins—amyloid beta-peptide protein in Alzheimer's disease and alpha-synuclein in Parkinson's disease. This chapter tells the tales of glutamate's roles in Alzheimer's and Parkinson's diseases as well as in two less common but no less terrible neurodegenerative disorders, Huntington's disease and amyotrophic lateral sclerosis.

ALZHEIMER'S DISEASE

One in every three people older than 65 will die with Alzheimer's disease. It is a devastating disease in which there occurs an inexorable decline in the ability to remember experiences. First, the ability to learn and remember new things is lost, and then old memories can no longer be recalled. This cognitive decline typically occurs over a period of 10 years or more, and most patients will require constant care for at least 5 years before they ultimately die. In 2019, more than 6 million people in the United States were living with a diagnosis of probable Alzheimer's disease at an annual medical and caregiving cost of more than $300 billion. By 2050, more than 15 million people in the United States will be living with Alzheimer's disease at an annual cost of more than $1 trillion.

Although so very many people suffer with Alzheimer's disease, their personal struggles are rarely communicated to the public, for an obvious and unfortunate reason. People with Alzheimer's progressively lose their ability to engage in substantive conversations because of their loss of short-term memory. They can read a book but will not remember what they read in the previous paragraph. They can converse but will not remember what was said only a minute ago. As the disease progresses, they will not be able to advocate for themselves, in contrast to people with most other diseases, who can advocate for themselves and can even establish their own foundations to raise money for research on their disease. However, in some instances a spouse or relative has helped increase the awareness of and research on Alzheimer's. For example, Ronald Reagan's wife, Nancy, worked with the Alzheimer's Association to raise millions of dollars for Alzheimer's disease research.

Because cognitive impairment can occur in disorders other than Alzheimer's disease, a definitive diagnosis of Alzheimer's disease can be established only after examination of a deceased person's brain tissue, where large amounts of amyloid plaques and neurofibrillary tangles are revealed in regions of the brain involved in learning and memory, such as the hippocampus, entorhinal cortex, parietal cortex, and frontal cortex. Amyloid plaques are extracellular accumulations of the protein amyloid beta-peptide, or Aβ (figure 8.1). Neurofibrillary tangles are twisted strands of polymers of the Tau protein that accumulate inside of neurons, which as a consequence degenerate and die. Neuroscientists have developed antibodies that bind selectively to either Aβ or Tau and thus can be used to visualize plaques and tangles in brain slices. The plaques and tangles in several microscope fields

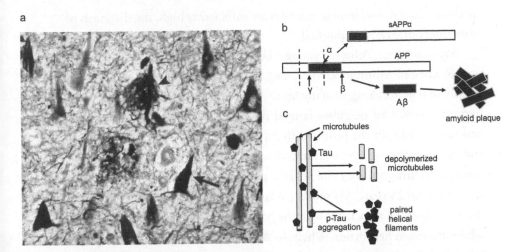

Figure 8.1

The histopathological hallmarks of Alzheimer's disease and the underlying molecular basis of their genesis. (a) Image of a tissue section from the brain of a person who died with Alzheimer's disease. The section was stained in a manner that reveals aggregated Tau filaments—neurofibrillary tangles—and amyloid plaques. The arrow points to an example of neurofibrillary tangles in a pyramidal neuron, and the arrowhead points to amyloid plaque. (b) Diagram showing the two main ways in which the beta-amyloid precursor protein (APP) is cleaved by enzymes. APP is an integral membrane protein, with one membrane-spanning domain. Aβ is a 40–42 amino acid sequence located partially within the cell membrane and partially in the extracellular space. Cleavage of APP by α-secretase (α) releases a large portion of APP called secreted APPα (sAPPα) from the cell. Sequential cleavages of APP by β-secretase (β) and γ-secretase (γ) releases Aβ from the cell. In Alzheimer's disease, Aβ self-aggregates outside of cells, forming amyloid plaques. (c) Tau is a protein that normally binds to and thereby stabilizes microtubules in the axons of neurons. In Alzheimer's disease, Tau becomes excessively phosphorylated (p-Tau) resulting from its detachment from microtubules. The microtubules depolymerize, and the p-Tau forms paired helical filaments inside neurons. These filamentous aggregations of p-Tau are called "neurofibrillary tangles."

are then counted, and if their numbers are sufficiently high, the diagnosis of Alzheimer's disease is established.

My older sister, Polly, my younger brother, Eric, and I grew up on a farm near Rochester, Minnesota. My father, DeWayne, had purchased the land mainly so that he, along with my brother and me, could train Standardbred harness horses. Our neighbor farmed the land—fields of alfalfa hay, corn, and oats—and split the profits with my father. Outside of school, training and racing the horses occupied considerable amounts of my time. My father was a prosecuting attorney and held the position of Olmsted county attorney for more than 25 years. My mother, Martha, had been a nurse at St. Mary's, the main hospital associated with the Mayo Clinic. She developed arthritis when she was in her forties. When she was 67, she had a knee joint replacement and shortly thereafter developed sepsis as a result of infection of the joint. She recovered from the infection but then had a stroke and died soon afterward. She was at high risk for a stroke because she had been a chain-smoker, was overweight, and had hypertension. At the time of her death, my father was in excellent health and fully self-sufficient.

During the next decade, my father seemed fine and was functioning well. But when he was in his late seventies, I noticed a subtle impairment of his short-term memory. During phone conversations, he would ask me the same question more than once because he had forgotten that he had already asked the question. Nevertheless, he was able to drive his car to the grocery store, the gas station, and a friend's house without difficulty, and he managed chores around the house and farm. However, a few years later his forgetfulness was resulting in problems in his management of his finances. My siblings and I decided to have him see a neurologist. Because my research had a heavy emphasis on brain aging and Alzheimer's disease, I was familiar with many neurologists who focus on Alzheimer's disease. I knew Ron Petersen at the Mayo Clinic, who studies patients with mild cognitive impairment or Alzheimer's disease. Petersen was the doctor for many famous people with dementia, including President Ronald Reagan. After testing my father's learning and memory abilities and evaluating MRI images of his brain, Petersen diagnosed him with "probable" Alzheimer's disease.

My brother, Eric, an equine veterinarian, was able to move in with my father and live with him for the next five years. Each year my father would return to the Mayo Clinic for follow-up evaluations. His learning and memory continued to deteriorate, and his hippocampi were progressively reduced in size. He died less than two months before his ninetieth birthday. Upon his death, his brain was removed, and pieces from several brain regions were taken and cut into small slices. The slices were then examined to determine how many Aβ plaques and Tau neurofibrillary tangles were present in those brain regions. It turned out that there was extensive loss of neurons in the hippocampus and frontal cortex but not enough Aβ plaques and tangles to meet the diagnosis of Alzheimer's disease. Neuropathologists have called this type of dementia "hippocampal sclerosis of aging." Studies have shown that about 20 percent of people diagnosed with probable Alzheimer's disease have a brain pathology like my father's.

The diagnosis of Alzheimer's as a unitary disease is semiarbitrary because it is based on the person having certain amounts of Aβ plaques and neurofibrillary tangles in their brain. But those amounts of plaques and tangles were established by a group of neurologists. In reality, people's brains exhibit a wide spectrum of plaque and tangle numbers that do not always correspond to specific cognitive deficits. Some people have very high numbers of amyloid plaques but little or no cognitive deficits. In contrast, my father is an example of someone with severe cognitive deficits but only a modest number of amyloid plaques. And some people will have a "vascular dementia," with little or no amyloid plaques and tangles but death of neurons in the gray matter and damage to axons in the white matter. There is also a rarer type of inherited disorder called "frontotemporal dementia" in which there are very large numbers of neurofibrillary tangles but no Aβ plaques. Frontotemporal dementia most commonly results from a mutation in the Tau protein. Others may suffer from dementia as a result of ministrokes.

In the late 1980s, it was thought that a major breakthrough had been achieved in Alzheimer's research when the gene for the Aβ precursor protein (APP) was identified (Kang et al. 1987; Weidemann et al. 1989). The US National Institutes of Health devoted large amounts of funding to neuroscientists to study how Aβ is produced from APP. Pharmaceutical companies invested billions of dollars in the development of drugs that would prevent the accumulation of Aβ or facilitate its removal from the brain. But over the three decades since then, all of the clinical trials of such "amyloid-centric" treatments have failed. In hindsight, this tunnel-vision view of Alzheimer's disease neglected other fundamental features of the disease. Indeed, several important findings suggested that amyloid is not the key to understanding Alzheimer's disease. For example, neuropathologists noticed that some older people who did not have dementia had large amounts of Aβ plaques in their brains, thus demonstrating that Aβ accumulation is not sufficient by itself to cause Alzheimer's disease. In addition, neurofibrillary tangles indistinguishable from those in Alzheimer's occur in people without Aβ plaques but with frontotemporal dementia. Moreover, neurons in the brains of mice genetically engineered to accumulate large amounts of Aβ in their brains do not degenerate, although the mice do exhibit some cognitive deficits. Finally, aging is the major risk factor for Alzheimer's, and research has identified several alterations that occur in the brain during aging that likely contribute to the demise of people with Alzheimer's (figure 7.1). These age-related alterations include impaired function of mitochondria; oxidative damage to proteins, DNA, and membranes; and the accumulation of molecular "garbage" in neurons.

Nevertheless, research on amyloid did provide insight into the normal function of APP and how abnormalities in its metabolism might contribute to many cases of Alzheimer's disease (Mattson 2004). Aβ is a 42-amino-acid peptide fragment of the much larger 695-amino-acid APP (figure 8.1). APP is a transmembrane protein: approximately half of Aβ is in the membrane, and the other half is outside the membrane. Aβ is liberated from APP by sequential enzymatic cleavages by the beta-secretase and gamma-secretase enzymes. Alternatively, APP can be cleaved in the middle of the Aβ sequence by the enzyme alpha-secretase. The latter cleavage liberates a large extracellular portion of APP called "secreted APP alpha," or sAPPα.

In rare instances, Alzheimer's disease is caused by mutations in the genes encoding APP or presenilin 1. Presenilin 1 is the enzyme gamma-secretase, which liberates Aβ from APP. The mutations in APP and presenilin 1 are inherited in a dominant manner such that a person with either a father or mother with a mutant gene has a 50 percent chance of having the mutation. People with such familial Alzheimer's disease mutations will develop symptoms at a relatively young age, usually when they are in their forties or fifties. Cultured cells and mice genetically engineered with human APP and/or presenilin 1 mutations exhibit altered enzymatic processing of APP, which results in increased Aβ production and reduced sAPPα production. In addition, findings from studies of cultured cells and transgenic mice with mutations in presenilin 1 suggest that the mutations alter Ca^{2+} release from the endoplasmic reticulum in a manner that renders neurons vulnerable to excitotoxicity (Bezprozvanny and Mattson 2008; Guo et al. 1999)

Studies of the effects of Aβ and sAPPα on brain cells have provided evidence that aberrant neuronal hyperexcitability occurs early in the disease process and causes neurofibrillary degeneration. One of my early discoveries in research on Alzheimer's was that healthy human neurons can withstand considerable amounts of Aβ, but when they are exposed to the Aβ, they become highly vulnerable to being damaged and killed by glutamate (Mattson, Cheng, et al. 1992). Further experiments revealed why Aβ renders neurons vulnerable to excitotoxicity. What happens is that Aβ accumulates on the surface of the neuron's membrane and causes oxygen free radicals to attack the fat molecules that form the cell membrane. More specifically, the free radicals attack the double bonds of unsaturated lipids, such as arachidonic acid. This process is called "lipid peroxidation" and is similar to what happens when fatty foods turn rancid. As a result of the membrane lipid peroxidation, small fragments of the lipid are released. One such fragment is called "4-hydroxynonenal," or HNE (figure 8.2).

Once liberated by the free radical attack on the membrane, HNE binds to and impairs the function of several proteins in the membrane that are critical for the neuron's ability to maintain its energy levels and ion gradients. These proteins include a glucose transporter that moves glucose into the neurons as well as the sodium and calcium pump proteins that move these ions out of the neuron to restore the charge potential across the membrane after the influx of the ions has been stimulated by glutamate. As a consequence, neurons may become hyperexcitable and so are prone to excitotoxic

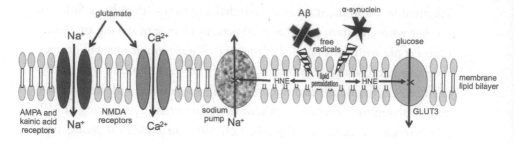

Figure 8.2

Illustration of how the aggregation of Aβ or alpha-synuclein may render neurons prone to excitotoxicity in Alzheimer's and Parkinson's diseases, respectively. As Aβ and alpha-synuclein aggregate on the membranes of neurons, they cause the production of free radicals, which attack double bonds in membrane lipids in a process called "lipid peroxidation." The lipid peroxidation liberates the nine-carbon lipid 4-hydroxynonenal, or HNE. The HNE can irreversibly bind to and thereby impair the function of neuron membrane proteins, including several that play important roles in protecting neurons against excitotoxicity. For example, studies have shown that by impairing the function of the membrane Na⁺ pump protein, HNE promotes membrane depolarization and exacerbates Ca²⁺ influx through NMDA receptor channels. HNE also impairs the function of the glucose transport protein GLUT3, resulting in an energy deficit, which further compromises the neuron's ability to constrain excitability.

degeneration. Indeed, findings from recent fMRI studies of human patients and mouse models of Alzheimer's disease suggest that neuronal network hyperexcitability occurs early in the disease.

Although increased production of Aβ has been a major focus of Alzheimer's disease research, the mutations in APP that cause early-onset inherited disease also result in reduced production of sAPPα. Research findings suggest that sAPPα plays important physiological roles in regulating neuronal excitability, synaptic plasticity, and stress resistance. In 1993, Bin Cheng, colleagues, and I discovered that sAPPα can protect cultured hippocampal neurons against glutamate excitotoxicity (Mattson, Cheng, et al., 1993). The mechanism involves a reduction in the accumulation of intracellular Ca²⁺. To understand how sAPPα protects neurons against excitotoxicity, Katsutoshi Furukawa recorded ion fluxes across the membrane of hippocampal neurons using patch-clamp electrodes. When he exposed the neurons to sAPPα, there occurred a rapid and reversible hyperpolarization of the membrane (Furukawa et al. 1996). Additional experiments showed that sAPPα activates a particular type of K$^+$ channel by a mechanism involving the second messenger cyclic GMP. Recent findings suggest that sAPPα can also reduce neuronal hyperexcitability by enhancing the activity of GABA receptors (Rice et al. 2019).

Increasing evidence suggests that sAPPα normally functions in the regulation of neuronal network activity, synaptic plasticity, and learning and memory. Mice in which the gene-encoding APP has been deleted exhibit impaired spatial learning and memory as well as a deficit in LTP at hippocampal synapses. Sabine Ring and colleagues found that when sAPPα is expressed in the APP-deficient mice, their learning and memory and LTP are restored to normal levels (S. Ring et al. 2007). Max Richter and colleagues reported similar results, showing that synapse numbers are reduced in APP-deficient mice and that treatment with sAPPα can restore synapse numbers (Richter et al. 2019). Another study reported that infusion of sAPPα into the hippocampus enhances LTP and improves learning and memory in very old rats (Xiong et al. 2017). Evidence suggests that these beneficial effects of sAPPα may be mediated by cyclic GMP and the transcription factor NF-κB (Barger and Mattson 1996).

Someone with Alzheimer's disease is more than 20 times as likely to have epileptic seizures than a person of the same age who does not have Alzheimer's disease. Changes that occur in the brain during normal aging may render neurons vulnerable to excitotoxicity (figure 7.1). Two such changes are oxidative stress and impaired neuronal energy metabolism. Working at the University of Kentucky, the neurologist Bill Markesbery found that levels of the membrane lipid peroxidation HNE are higher in the brains of people with Alzheimer's disease than in people of the same age without Alzheimer's (Williams et al. 2006). The study showed that HNE levels are also elevated in older people who have relatively mild cognitive impairment. Therefore, increased free radical attack on membrane lipids is a common feature of Alzheimer's disease and likely contributes to impaired function of the Na$^+$ and Ca^{2+} pump proteins, thereby rendering neurons vulnerable to excitotoxicity (Mattson 2004).

Neurons' ability to acquire and utilize energy from dietary carbohydrates is compromised in people with mild cognitive impairment and Alzheimer's disease (Cunnane et al. 2020). This cellular energy deficit results from impaired function of the neuronal membrane glucose transporter. In addition, there is considerable evidence that neurons suffer from impaired mitochondrial function to some extent during normal aging and much more so in Alzheimer's disease (Lanzillotta et al. 2019). Chapters 6 and 7 described how reduced ATP levels caused by mitochondrial toxins or stroke makes neurons prone to hyperexcitability. ATP depletion is also thought to occur in neurons affected in Alzheimer's disease.

Neuropathologists' examination of the brains of people who died with Alzheimer's disease have shown that GABAergic inhibitory neurons degenerate in the early stages of the disease (Mattson 2020). Because GABAergic inhibitory neurons function to prevent aberrant activity in neuronal networks, their early demise likely contributes to unconstrained glutamatergic activity and consequent degeneration of glutamatergic neurons. The GABAergic neurons are vulnerable to excitotoxicity because of their high firing rate, which stems from repetitive stimulation by the glutamatergic neurons that innervate them. As a consequence of the rapid firing, the GABAergic neurons produce very high amounts of oxygen free radicals and require relatively large amounts of energy.

Further evidence that neuronal networks become hyperexcitable early in the course of Alzheimer's disease comes from studies of mice genetically engineered to accumulate large amounts of Aβ in their brains. EEG recordings of brain neuronal network activity have shown that these mice are prone to developing epileptic seizures as they age. When these mice are bred with mice lacking the mitochondrial protein SIRT3, the offspring develop severe seizures and die from the seizures when they are young (A. Cheng, Wang, et al. 2020). This occurs even in mice with only a 50 percent reduction in SIRT3 levels. Mice lacking SIRT3 but also with no Aβ in their brains do not develop seizures. Therefore, whereas mild impairment of mitochondrial function alone does not cause seizures, it greatly exacerbates seizures when amyloid accumulates in the brain.

Findings from cell cultures, animal modeling, and human studies suggest that hyperactivity of glutamatergic neurons is both sufficient and necessary to cause neurofibrillary degeneration. Early evidence for this came from a study in 1990 in which I showed that exposure of cultured hippocampal neurons to levels of glutamate that cause a sustained elevation of intracellular Ca^{2+} levels results in changes in Tau similar to those in neurofibrillary tangles in the hippocampus of Alzheimer's patients (Mattson 1990). Bin Cheng and I found that glucose deprivation that models the impaired neuronal glucose transport in Alzheimer's causes similar changes in Tau (B. Cheng and Mattson 1992a). Moreover, Elicia Elliot, Robert Sapolsky, and their colleagues reported that induction of seizures with kainic acid causes neurofibrillary-like degeneration of hippocampal pyramidal neurons in rats (Elliot et al. 1993).

Functional MRI studies in transgenic mice that develop both Aβ and Tau pathologies demonstrated a strong association between local hyperactivity of glutamatergic circuits and Tau pathology (D. Liu, Lu, et al. 2018). Holly Hunsberger and colleagues found that riluzole, a drug that enhances the removal of glutamate from synapses, prevents Tau pathology and cognitive decline in a mouse model of frontotemporal dementia (Hunsberger et al. 2015). Using cell culture and optogenetic stimulation of neurons in vivo, Jessica Wu, Karen Duff, and their coworkers showed that pathological variants of Tau can be released from cells and taken up by adjacent cells (J. Wu et al. 2016). When they increased the activity of glutamatergic neurons, the disease-causing Tau was released from those neurons and transferred trans-synaptically to other neurons. In addition, patients with severe epilepsy exhibit Tau pathology in the temporal lobe, as in Alzheimer's disease, and the Tau pathology is correlated with cognitive decline (Tai et al. 2016). Therefore, neurofibrillary degeneration can occur as a direct result of hyperexcitability of glutamatergic neurons in humans.

But why, then, are some older people with high amounts of Aβ in their brains still "sharp as a tack"? There presumably are factors that protect neurons from being damaged and killed by Aβ. One such factor may be the amount of the neurotrophic protein BDNF present in the brain during aging. BDNF levels decrease in the brain during normal aging and much more so in Alzheimer's disease, and studies have shown BDNF can protect neurons from being damaged and killed by Aβ (Arancibia et al. 2008). BDNF can also protect neurons against excitotoxicity. Other factors that may protect the aging brain against Alzheimer's disease are discussed in chapter 11 and include robust GABAergic neurons, mitochondrial stress resistance, and well-functioning autophagy, antioxidant defenses, and molecular-repair mechanisms.

Aβ causes oxidative stress, which can render neurons vulnerable to excitotoxicity (Mattson, Cheng, et al. 1992). It might therefore be the case that some older people's neurons can withstand oxidative stress even in the face of Aβ accumulation. This possibility is supported by studies showing that neurons can be protected from higher amounts of Aβ when their antioxidant defenses and mitochondrial resilience are bolstered. For example, hippocampal neurons are resistant to being killed by Aβ when they are treated with glutathione (Mark, Lovell, et al. 1997). Glutathione is normally produced by neurons and functions as an antioxidant that binds to and thereby neutralizes the toxic membrane lipid peroxidation product HNE. Glutathione levels are reduced in the brains of people with Alzheimer's disease and to a lesser extent in older people with mild cognitive impairment compared to in older people whose cognition is normal. Glutathione levels might be higher in the brains of older people with large amounts of amyloid plaques but no cognitive impairment, although this remains to be determined.

Some people may be blessed with well-functioning mitochondria in their neurons as they age. Hundreds of published studies have provided evidence that mitochondria are adversely affected by aging and Alzheimer's disease. Their dysfunction likely results in a cellular energy deficit that puts neurons at high risk for excitotoxicity. Scientists are working to develop treatments for Alzheimer's disease that bolster mitochondrial function. Yuyoung Hou and Vilhelm Bohr at the National Institute on Aging developed a mouse model that exhibits many of the core features of Alzheimer's disease, including Aβ plaques, neurofibrillary tangles, neuronal death, and cognitive impairment. These Alzheimer's mice have a reduced amount of nicotinamide adenine dinucleotide (NAD+) in their brains. NAD+ is important for mitochondrial energy production and increases the activity of SIRT3, which, as described in the section on epilepsy in chapter 7, can protect neurons against excitotoxicity. To bolster mitochondrial function, the Alzheimer's mice were treated with nicotinamide riboside, a chemical precursor of NAD+. The treatment did not reduce the amount of Aβ accumulation in the brain but did reduce the number of neurofibrillary tangles and prevented death of neurons (Hou et al. 2021). Nicotinamide riboside treatment increased SIRT3 activity and ameliorated cognitive deficits. Altogether, the results of the study demonstrated that it is possible to keep neurons alive and functioning well even in the face of high amounts of Aβ plaque in the brain.

What is the potential for interventions that protect against glutamatergic hyperactivity in the prevention and treatment of Alzheimer's disease? As of now, the only drug approved by the US Food and Drug Administration that may slow the course of the disease is memantine (McShane et al. 2019). This drug acts by blocking NMDA receptor channels after they open in response to membrane depolarization. However, the benefit for people with Alzheimer's is modest. Bimonthly treatment with ketamine, another drug that is an NMDA receptor open-channel blocker, is effective in treating major depression but has not been evaluated in people with Alzheimer's disease.

Given that seizures are common in Alzheimer's disease, drugs used to treat epilepsy are sometimes prescribed for those who have Alzheimer's. Emerging evidence suggests a potential for such drugs to slow the course of the disease. For example, in one study levetiracetam treatment improved cognition and normalized brain EEG activity in people with mild to moderate Alzheimer's disease (Musaeus et al. 2017). Another study showed that people with mild cognitive impairment exhibit hyperactivity of hippocampal neurons and that two weeks of treatment with levetiracetam normalized the activity. Importantly, spatial cognition was significantly improved in people treated with levetiracetam compared to in people receiving a placebo (Bakker et al. 2012). Levetiracetam is thought to reduce neuronal network excitability by inhibiting presynaptic voltage-dependent Ca^{2+} channels, thereby reducing the release of glutamate.

Hypertension during midlife is a risk factor for Alzheimer's disease. People whose hypertension is controlled with drugs reduce their risk for Alzheimer's compared to those whose hypertension is not controlled. One drug that is very effective in reducing blood pressure is diazoxide. This drug reduces blood pressure by opening K^+ channels in vascular smooth muscle cells, resulting in their relaxation. Diazoxide also activates K⁺ channels in the outer membrane of neurons, thereby reducing their excitability. Interestingly, diazoxide also activates K⁺ channels in the mitochondrial inner membrane, resulting in the neurons' increased resistance to metabolic and oxidative stress. One study showed that treatment with a low dose of diazoxide ameliorates cognitive deficits and lessens Aβ and Tau pathologies in transgenic mice with APP, presenilin 1, and Tau mutations (D. Liu, Pitta, et al. 2010). Thus far, there have been no clinical trials of low-dose diazoxide in people with Alzheimer's.

Recent findings suggest that ketones might prevent neuronal degeneration and cognitive decline in Alzheimer's disease. One study showed that dietary supplementation with a ketone ester lessens the impact of the disease process in the triple-transgenic mouse model of Alzheimer's disease (3xTgAD) (Kashiwaya et al. 2013). The ketone beta-hydroxybutyrate is the same ketone produced from fats in the body during fasting. The mice were fed either their normal diet or a diet containing the ketone ester. Their learning and memory abilities were measured at four and seven months after diet initiation. The 3xTgAD mice in the ketone ester group had significantly better performance in the cognitive tests compared to those in the control diet group. Moreover, the ketone ester reduced anxiety-like behaviors. Aiwu Cheng, Ruiqian Wan, and their collaborators pursued experiments to understand how the ketone ester might protect neurons against Alzheimer's disease (Cheng, Wang, et al. 2020). Their findings suggest that the ketones act by bolstering mitochondrial function.

Findings from studies of humans support a potential benefit of intermittent fasting and the ketone ester in people at risk for or with mild cognitive impairment and Alzheimer's disease. First, people with obesity and/or diabetes are at increased risk for Alzheimer's disease, and intermittent fasting can be very effective in preventing and even reversing both obesity and diabetes (Mattson 2022). Second, the Canadian neuroscientist Steve Cunnane has used positron emission tomography (PET), a brain-imaging method, to measure the relative abilities of neurons in the brain to use either glucose or ketones as their energy source (Cunnane et al. 2020). He found that when people eat a typical diet with considerable carbohydrates, their brain cells primarily use glucose as an energy source. In contrast, during fasting or when people are on a ketogenic diet with no carbohydrates, the brain cells primarily use ketones as their energy source. Additional studies showed that brain cells' ability to utilize glucose is reduced in people with mild cognitive impairment, but the cells remain able to utilize ketones (Neth et al. 2020). When taken together with the data from the animal studies described in the preceding paragraph, Cunnane's findings suggest that people who are at risk for or in the early stages of Alzheimer's disease might benefit from intermittent fasting and a ketone ester.

It would certainly be wonderful if there were drugs that could reverse or even halt the progression of cognitive impairment in Alzheimer's disease. However, given that many neurons have already died by the time someone is diagnosed, this seems unlikely. In contrast, interventions that reduce one's risk for developing Alzheimer's as they age are a reality. Chapter 11 describes the evidence that lifelong exercise, intellectual challenges, and moderation in energy intake are three such hedges against Alzheimer's disease as well as stroke and Parkinson's disease.

PARKINSON'S DISEASE

More than one million people in the United States are currently living with Parkinson's disease. Someone with this disease experiences progressively worsening difficulty in properly controlling the movement of their limbs. Shakiness of the hands is often the first problem noticed by the patient, although they may also have some difficulty walking and muscle stiffness. These so-called motor manifestations of Parkinson's disease are a consequence of the degeneration of neurons in the substantia nigra, located in the upper part of the brainstem. The neurons in the substantia nigra that degenerate use the neurotransmitter dopamine, and their activity normally prevents unwanted body movements. During the early stages of the disease, many dopaminergic neurons remain alive but are unable to produce dopamine. The motor symptoms can be relieved by giving levodopa to people with Parkinson's, a molecule that dopaminergic neurons can convert directly into dopamine. However, levodopa does not slow the degeneration and death of the neurons.

As with Alzheimer's disease, most cases of Parkinson's disease occur late in life and have no known genetic cause. The actor and science enthusiast Alan Alda, the evangelist Billy Graham, and the singer Linda Ronstadt are examples of people diagnosed with Parkinson's when they were in their sixth, seventh, or eighth decade of life (although Ronstadt's diagnosis was later changed to progressive supranuclear palsy). One risk factor for Parkinson's disease is a history of head trauma. There is also evidence that repeated exposures to certain pesticides, such as rotenone, can increase one's risk for Parkinson's. In chapter 6, you learned about natural and man-made neurotoxins that damage and kill neurons by excitotoxicity. Some of those toxins, such as kainic acid and domoic acid, directly activate glutamate receptors excessively. Others, such as the pesticides rotenone and paraquat, cause excitotoxicity indirectly by impairing the function of mitochondria in neurons. Because of their high firing rates and energy requirements, dopaminergic neurons are particularly vulnerable to such mitochondrial toxins.

The actor Michael J. Fox was diagnosed with Parkinson's at the age of 29. Developing Parkinson's at such a young age is uncommon and is usually caused by an inherited gene mutation. From the public record, it is not clear whether one of his parents had early-onset Parkinson's. According to Wikipedia, Fox believes he may have been exposed to a chemical that caused his Parkinson's disease: "I used to go fishing in a river near paper mills and eat the salmon I caught; I've been to a lot of farms; I smoked a lot of pot in high school when the government was poisoning the crops. But you can drive yourself crazy trying to figure it out."

Approximately 5 percent of Parkinson's disease cases are inherited. People with inherited Parkinson's disease typically begin to exhibit symptoms when they are in their thirties, forties, or fifties. In 1997, geneticists at the US National Institutes of Health reported the identification of a gene mutation associated with inherited Parkinson's disease in an Italian family (Polymeropoulos et al. 1997). The mutation is inherited in an autosomal dominant manner such that if either the father or mother has the mutation, then their child has a 50 percent chance of also having the mutation. The identified gene codes for the protein alpha-synuclein, which is produced mainly by neurons and is believed to play a role in regulating the release of glutamate at synapses. However, the mutant form of alpha-synuclein has a propensity to self-aggregate inside of neurons and may in this way clog up the neuron's molecular garbage-disposal system. As a result, damaged mitochondria accumulate, and the neuron suffers from an energy deficit that can increase its vulnerability to excitotoxicity. Interestingly, in one family with inherited Parkinson's, the affected individuals have a triplication of the alpha-synuclein gene but no change in the gene's DNA sequence (Singleton et al. 2003). This shows that increasing the amount of alpha-synuclein in neurons by just 33 percent is sufficient to cause Parkinson's disease.

Like Aβ, alpha-synuclein is prone to self-aggregation even in the absence of mutations. Also like Aβ, alpha-synuclein oligomers form and accumulate on cell membranes and cause membrane lipid peroxidation, which renders the neurons vulnerable to excitotoxicity. Examination of the brains of people who died with Parkinson's disease revealed that levels of the lipid peroxidation product HNE are greatest in neurons with high amounts of alpha-synuclein accumulation. HNE can cause the aggregation of alpha-synuclein, resulting in further accumulation of alpha-synuclein in the neurons (Qin et al. 2007).

Evidence suggests that alpha-synuclein oligomers can be transferred from one neuron to another, most likely at synapses. In one study, Shi Zhang and colleagues discovered that alpha-synuclein oligomers accumulate in tiny membranous vesicles called "exosomes" (Zhang et al. 2018). When neurons are exposed to HNE they release the exosomes. The exosomes can then fuse with the membrane of adjacent neurons, resulting in the accumulation of alpha-synuclein oligomers in those neurons. When exosomes containing alpha-synuclein oligomers are injected into the brains of mice, alpha-synuclein pathology spreads to anatomically connected brain regions. This mechanism may explain how alpha-synuclein pathology spreads across neuronal networks.

Dorit Trudler, Stuart Lipton, and their coworkers found that alpha-synuclein oligomers can cause the release of glutamate from astrocytes and can enhance the activation of NMDA receptors by glutamate (Trudler et al. 2021). Sandra Huls and coworkers found that glutamatergic synaptic activity is increased by alpha-synuclein oligomers (Huls et al. 2011). As with Aβ, these effects of alpha-synuclein on glutamatergic synapses may result from an enhanced depolarization of the cell membrane owing to impairment of the Na^+ pump protein by the lipid peroxidation product HNE (figure 8.2). In addition, HNE can bind to the membrane glucose transport protein GLUT3, thereby diminishing the amount of glucose available for ATP production.

Soon after the discovery of the alpha-synuclein gene mutation, geneticists sequenced DNA from people in other families with inherited Parkinson's and identified mutations in different genes in some of those families, including genes that encode the proteins Parkin, deglycase DJ-1, leucine-rich repeat kinase 2 (LRRK2), and PTEN-induced kinase 1 (PINK1; PTEN stands for "phosphatase and tension homolog"). These proteins were then shown to normally serve the functions of increasing the resistance of mitochondria to stress, removing damaged proteins and mitochondria, and removing excessive amounts of free radicals. Studies of cultured brain cells and transgenic mice with mutations in the proteins Parkin, DJ-1, LRRK2, and PINK1 have shown that the mutations perturb glutamatergic neurotransmission and render neurons vulnerable to excitotoxicity (van der Vlag, Havekes, and Heckman 2020).

Mutations in the genes encoding Parkin, PINK1, and DJ-1 are inherited in an autosomal recessive manner such that someone with Parkinson's caused by such mutations has inherited one dysfunctional gene from each parent. They are "loss of function" mutations. The evidence that Parkin, PINK1, and DJ-1 normally protect neuronal networks against hyperexcitability and excitotoxicity is strong. Parkin is a ubiquitin ligase—an enzyme that adds a small protein called "ubiquitin" to other proteins. By this mechanism, Parkin tags damaged proteins for removal by autophagy and enzymatic degradation in the proteasome. Glutamatergic neurons in the brains of Parkin-deficient mice exhibit increased sensitivity to kainic acid and are prone to excitotoxicity. Overproduction of normal Parkin in postsynaptic neurons reduces excitatory synaptic transmission, whereas removal of Parkin increases synaptic excitation. PINK1 is a kinase that phosphorylates proteins involved in mitochondrial bioenergetics, Ca^{2+} handling, and mitophagy. Mutations of PINK1 that cause Parkinson's disease compromise these functions. Studies have shown that the activities of neurons in several brain regions are increased in rats or mice with a genetic PINK1 deficiency. Midbrain dopaminergic neurons in PINK1-deficient mice are hyperexcitable and exhibit enhanced spontaneous burst firing (Bishop et al. 2010). DJ-1 apparently functions as an antioxidant protein and by this mechanism can prevent the aggregation of alpha-synuclein. By reducing oxidative stress, DJ-1 may also protect neurons against excitotoxicity.

LRRK2 mutations are responsible for most cases of inherited Parkinson's disease. The mutations are inherited in an autosomal-dominant manner. LRRK2 is a kinase that interacts with Parkin as well as with several proteins on the outer mitochondrial membrane. Young mice expressing mutant LRRK2 exhibit increased activity at glutamatergic and dopaminergic synapses on striatal medium spiny neurons, with a subsequent deficit in dopaminergic neurotransmission as the mice age. The early increase in glutamatergic neurotransmission is followed by degeneration of dendrites and in some animal models by neuronal death (Plowey et al. 2014). Several mechanisms have been proposed to explain the excitotoxic dendritic atrophy and cell death caused by LRRK2 mutations, including impairments of mitochondrial function and autophagy.

A recent remarkable twist in understanding Parkinson's disease came from evidence suggesting that the neurodegenerative process may actually begin in the gut and not the brain (Del Tredici and Braak 2016)! Using an antibody that binds to alpha-synuclein, the German neuropathologist Heiko Braak worked to understand which neurons are first affected in Parkinson's disease. Braak used a method called "immunohistochemistry" to visualize the location and relative amount of alpha-synuclein in tissue sections from different brain regions of patients that died at different stages of Parkinson's disease. Surprisingly, he found that the substantia nigra dopaminergic neurons were not the first neurons in the brain to exhibit alpha-synuclein accumulation: the first were neurons in the brainstem that use the neurotransmitter acetylcholine. Those cholinergic neurons are part of the parasympathetic nervous system and their axons pass through the vagus nerve and form synapses on cells in the heart, gut, and other organs. Activation of the brainstem parasympathetic neurons slows heart rate and increases gut motility.

The axons in the vagus nerve innervate neurons adjacent to the intestines. The latter neurons are part of the so-called enteric nervous system and form synapses on smooth muscle cells that encircle the intestinal wall. Their activation causes the muscle cells to contract and so to squeeze food through the intestine. When Braak examined tissue from the intestines of deceased Parkinson's patients, he found that there was robust accumulation of alpha-synuclein in the neurons that stimulate gut motility. This was true even in patients who had just been diagnosed with Parkinson's and then died shortly thereafter from an unrelated cause such as a heart attack. Braak published his findings in 2006 and proposed what was at the time considered a radical idea:

> Alpha-synuclein immunoreactive inclusions were found in neurons of the submucosal Meissner plexus, whose axons project into the gastric mucosa and terminate in direct proximity to fundic glands. These elements could provide the first link in an uninterrupted series of susceptible neurons that extend from the enteric to the central nervous system. The existence of such an unbroken neuronal chain lends support to the hypothesis that a putative environmental pathogen capable of passing the gastric epithelial lining might induce alpha-synuclein misfolding and aggregation in specific cell types of the submucosal plexus and reach the brain via a consecutive series of projection neurons. (Braak et al. 2006, 67)

Many Parkinson's researchers ignored Braak's finding because the pathology in the brain, not the gut, is responsible for the motor symptoms of Parkinson's disease. However, there were tantalizing clinical data to support his gut-to-brain hypothesis. Reviews of the clinical history of patients suggested that most suffer from chronic constipation before they are diagnosed with Parkinson's disease. This is by no means convincing evidence that the neurodegenerative process is propagated from the gut to the brain. However, in 2015 Danish investigators reported that patients who had their vagus nerve cut just above the gut as a treatment for severe gastric ulcers have a reduced risk of developing Parkinson's disease (Svensson et al. 2015). Then in 2019, Sangjune Kim and coworkers showed that injection of aggregating alpha-synuclein into the gut muscles of mice results in spreading of alpha-synuclein pathology to the brain, with a progression from the dorsal motor nucleus of the vagus to the raphe nucleus to the substantia nigra (S. Kim et al. 2019). As a result, the dopaminergic neurons degenerate, and the mice exhibit an impaired ability to control their body movements.

Evidence suggests that chronic gut inflammation can accelerate the progression of Parkinson's disease. Several population-based studies have shown that people with inflammatory bowel disorders are at increased risk for developing Parkinson's disease as they age. Geneticists have shown that people with mutations of LRRK2 are at increased risk for inflammatory bowel disorders, suggesting similarities in the cellular alterations in these disorders and Parkinson's (Villumsen et al. 2019). Working in my laboratory, Yuki Kishimoto showed that chronic mild gut inflammation accelerates the development of alpha-synuclein pathology in the gut and brain and hastens the onset of dopaminergic neuron degeneration and associated motor symptoms in a mouse model of Parkinson's (Kishimoto, Zhu, et al. 2019). Inflammation was increased in the gut and brain but not in the blood, which is consistent with the notion of retrograde propagation of the disease from the gut to the brain via the vagus nerve.

Recent findings suggest that the composition of gut bacterial species—microbiota—can influence Parkinson's disease. Timothy Sampson and colleagues found that when they transplanted gut microbiota from people with Parkinson's into the gut of alpha-synuclein mutant mice, brain inflammation and motor deficits were exacerbated, but not when they transplanted microbiota from people without Parkinson's (Sampson et al. 2016). It is not yet known whether gut inflammation increases the sensitivity of brainstem vagal neurons and midbrain dopaminergic neurons to glutamate. However, this connection seems likely because the gut inflammation increases alpha-synuclein accumulation in those neurons, and studies have shown that such accumulations of alpha-synuclein do increase the sensitivity of neurons to glutamate.

Many studies of animal models of Parkinson's disease have aimed to identify treatments that can protect dopaminergic neurons from being damaged and killed. One approach is to use drugs that block glutamate receptors. Animal studies have shown that drugs that block NMDA receptors can protect dopaminergic neurons against MPTP (Brouillet and Beal 1993). Dopaminergic neurons can also be protected from being damaged by MPTP in mice with a drug that blocks kainic acid receptors (Stayte et al. 2020). But problems arise in the use of glutamate receptor blockers to treat people with Parkinson's. Such drugs can have serious side effects, including cognitive impairment and hallucinations, particularly when they are administered over long periods because they impair the function of glutamatergic synapses that are required for learning and memory, decision-making, and other cognitive processes.

It has been found that two neurotrophic factors—BDNF and glial cell–derived neurotrophic factor (GDNF)—can protect dopaminergic neurons against degeneration in experimental models of Parkinson's disease. Each neurotrophic factor can also protect neurons against direct excitotoxicity (B. Cheng and Mattson 1994; Emerich et al. 2019). Several methods have been used to deliver these neurotrophic factors to the striatum or substantia nigra: infusion using a minipump implanted under the skin and transplantation of cells expressing high levels of the neurotrophic factors. In one clinical trial, GDNF was infused into the putamen—part of the basal ganglia where dopaminergic neurons' axon terminals reside—in 10 people with Parkinson's disease for six months. During the period of infusion and for up to nine months thereafter, the 10 participants' motor symptoms improved significantly (Slevin et al. 2007). However, in a more recent trial GDNF infusion into the putamen showed no significant clinical benefit in people with Parkinson's (Whone et al. 2019).

MPTP, rotenone, and gene mutations that cause early-onset Parkinson's kill neurons by impairing the function of their mitochondria and causing free radical production. Therefore, one approach to protecting the neurons is to treat them with antioxidants or chemicals that bolster mitochondrial function. The chemical coenzyme Q10 seemed promising in animal models of Parkinson's. Coenzyme Q10 is normally present in the membrane of mitochondria, where it plays a role in electron transport and is also an antioxidant. Flint Beal and colleagues found that when mice were treated with coenzyme Q10 and nicotinamide, their dopaminergic neurons were more resistant to being damaged by MPTP (Beal et al. 1998). Unfortunately, several trials in patients with Parkinson's disease failed to show a benefit of coenzyme Q10. Similarly, despite showing promise in MPTP-treated laboratory animals, creatine, a natural chemical that can increase ATP levels in cells, also failed to show a benefit for Parkinson's patients.

An approach that has shown promise in slowing the progression of Parkinson's disease in recent clinical trials has its origins in research on diabetes. The hormone glucagon-like peptide-1 (GLP-1) is produced by cells in the wall of the intestines. When food enters the intestines, GLP-1 is secreted into the blood. GLP-1 then functions to lower blood glucose levels in two ways: it increases the sensitivity of muscle, liver, and other cells to insulin, and it stimulates insulin release from cells in the pancreas. In addition, GLP-1 inhibits appetite, thereby preventing overeating. However, after entering the blood, GLP-1 circulates for only a few minutes because it is degraded by the enzyme dipeptidyl-peptidase 4 (DPP4). Working at the National Institute on Aging in Baltimore, the endocrinologist Josephine Egan and the pharmacologist Nigel Greig teamed up and modified GLP-1 in a way that makes it resistant to being degraded by DPP4 and so circulates in the blood for many hours. Egan and Greig named this stable form of GLP-1 "Exendin-4" and showed that it was very effective in controlling and even reversing diabetes in laboratory animals and humans. Exendin-4 is now widely prescribed for patients with type 2 diabetes.

But the GLP-1 story did not end with diabetes. Nigel Greig had found that cultured neural cells can respond to GLP-1. He and I then collaborated on studies that showed that GLP-1 and Exendin-4 can protect hippocampal neurons against excitotoxicity (Perry et al. 2002). We found that when neurons were treated with GLP-1, the elevation of intracellular Ca^{2+} levels caused by glutamate was attenuated. In subsequent studies, Exendin-4 increased the resistance of dopaminergic neurons in mice to degeneration caused by MPTP (Y. Li et al. 2009). These promising results in animal studies prompted a randomized controlled trial of Exendin-4 in Parkinson's patients in England. The trial, which was headed by the neurologist Tom Foltynie, showed that Exendin-4 significantly slows the worsening of symptoms during a one-year period (Athauda et al. 2017).

Bolstering neuronal energy with ketones is another promising approach for improving symptoms and slowing the disease process in Parkinson's. Impaired ability to exercise is one consequence of Parkinson's, and, indeed, a recent study showed that a dietary ketone ester can improve cycling endurance in patients with Parkinson's disease (Norwitz et al. 2020). As described in the preceding section, the same ketone ester has been shown to prevent neuronal network hyperexcitability in an animal model of Alzheimer's disease.

Finally, it may be possible to protect neurons and preserve their function in Parkinson's patients using chemicals that engage adaptive cellular stress responses in neurons. This approach is based on hormesis—the evolutionarily conserved mechanisms by which cells respond to a mild and transient stress in ways that increase the cells' resistance to ongoing or subsequent stress. Animal studies have shown that two chemicals, 2-deoxyglucose and 2,4-dinitrophenol (DNP), can activate some adaptive cellular responses. When animals are treated with 2-deoxyglucose, it is transported into cells throughout the body and brain, just as glucose is. But in contrast to glucose, 2-deoxyglucose cannot be used to produce ATP and in fact prevents glucose from being used for ATP production. In this way, 2-deoxyglucose mimics the effects of fasting, and, indeed, fats are mobilized, and ketones are instead used as an energy source for cells in animals or humans administered 2-deoxyglucose. When mice are treated with 2-deoxyglucose, their dopaminergic neurons become relatively resistant to being damaged by MPTP (Duan and Mattson 1999). Low doses of DNP also protected dopaminergic neurons and improved functional outcome in mice with an alpha-synuclein mutation (Kishimoto, Johnson, et al. 2020). DNP imposes a mild metabolic stress by causing a proton (H^+) leak across the mitochondrial inner membrane. Both 2-deoxyglucose and DNP can induce the expression of BDNF, which may contribute to their neuroprotective effects.

The safest ways to engage adaptive stress responses in cells of the body and brain, however, are not with drugs but with exercise and intermittent fasting. Chapter 11 describes evidence that these lifestyle factors may reduce the risk of Parkinson's disease and may also benefit people who already have this neurodegenerative disorder.

HUNTINGTON'S DISEASE

Woody Guthrie was a folk music songwriter and singer whose most famous song is "This Land Is Your Land." Guthrie grew up in Oklahoma. When he was 14 years old, his mother was hospitalized with Huntington's disease, a fatal inherited neurological disorder. Woody in turn inherited from her a copy of the mutated gene that caused Huntington's. The symptoms of Huntington's include psychiatric problems, involuntary writhing movements of the limbs known as "chorea," and difficulty speaking clearly. Mood swings, difficulty concentrating, and memory lapses are also features of the disease. These symptoms will typically begin to emerge when the affected person is in their thirties or forties. In 1952 at the age of 40, Guthrie was committed to a psychiatric institute because of extreme mood swings that included bouts of rage. He was released and managed to function fairly well for several years, but by 1965 his disease had progressed to the point that he could no longer speak. He died at the age of 55.

Huntington's disease is rare, with about one in 20,000 people in the United States affected. However, because it is inherited in a dominant manner, a child born to a parent with Huntington's disease has a 50 percent chance of inheriting the disease from that parent. There are clusters of families throughout the world in which Huntington's disease is common. Moreover, the clusters are much more common in families of western European descent than in families of Asian or African descent. The disease is caused by a mutation in the gene that encodes the protein huntingtin. The mutation is unusual in that it involves the insertion of long repeats of the three-base DNA sequence cytosine–adenine–guanine (CAG) in the gene. The CAG codes for the amino acid glutamine, and so the abnormal huntingtin protein has a long string of glutamines.

The neurons affected most harshly in Huntington's disease are the medium spiny neurons in the basal ganglia. The medium spiny neurons are large GABAergic neurons that receive input from the dopaminergic neurons in the substantia nigra. The function of the medium spiny neurons that degenerate in Huntington's is to inhibit unwanted body movements, and so when they degenerate, the person will exhibit continuous body movements. Two features shared by the GABAergic medium spiny neurons of the striatum that degenerate in Huntington's and the dopaminergic neurons of the substantia nigra that degenerate in Parkinson's are that they have high amounts of NMDA receptors and are highly active. They fire action potentials at a higher frequency than most other neurons. They also have greater numbers of mitochondria, which are necessary to provide the energy to support their high activity levels. Their high energy demand is thought to make medium spiny neurons and dopaminergic neurons vulnerable to excitotoxicity.

Evidence suggests that mutant huntingtin compromises mitochondrial function (Chang et al. 2006). Like alpha-synuclein in Parkinson's disease, mutant huntingtin binds to itself and aggregates within neurons. This accumulation of the abnormal huntingtin damages mitochondria. Moreover, the neurons' ability to remove damaged mitochondria is impaired. As a result, the neurons accumulate dysfunctional mitochondria that produce little ATP and instead produce high amounts of free radicals. The reduced energy levels and the oxidative stress caused by the free radicals impair the neurons' ability to pump Na^+ and Ca^{2+} out after they fire. In this way, the medium spiny neurons are highly vulnerable to excitotoxicity.

In addition to its adverse effects on mitochondria, mutant huntingtin may predispose neurons to excitotoxicity in more specific ways. One way is by increasing the number of NMDA receptors in the neurons' membrane. Sonia Marco and colleagues showed that mutant huntingtin causes intracellular NMDA receptors to be inserted in the plasma membrane (Marco et al. 2013). Using molecular genetic methods, they found that increasing the number of NMDA receptors causes striatal neuron degeneration in mice with normal huntingtin. Deletion of NMDA receptors in mutant huntingtin mice reduces degeneration of striatal neurons and ameliorates motor and cognitive deficits.

Postmortem analyses have demonstrated reduced levels of BDNF in the striatum and cerebral cortex of people who died with Huntington's disease. Similarly, BDNF levels are reduced in the striatum and cerebral cortex in transgenic mice expressing mutant human huntingtin. Evidence suggests that mutant huntingtin can inhibit the transcription of the gene encoding BDNF (Zuccato et al. 2003). Increasing the amount of BDNF in the brains of mutant huntingtin mice, either by gene therapy or by infusion of BDNF, reduces degeneration of striatal and cortical neurons and ameliorates motor deficits. Because BDNF can protect neurons against excitotoxicity, it is likely that a BDNF deficit plays a role in Huntington's disease.

A wide range of approaches have been reported effective in slowing disease progression in mice with mutant huntingtin (Crook and Housman, 2011), including NMDA receptor antagonists and antioxidants. Other treatments effective in animal models of Huntington's disease include antidepressants such as paroxetine, the GLP-1 agonist Exendin-4, the mitochondrial uncoupler DNP, and electroconvulsive therapy (Duan, Guo, et al. 2004; B. Martin et al. 2009; Mughal et al. 2011; B. Wu et al. 2017). What these treatments have in common is that they all boost BDNF levels in the brain. Infusion of BDNF or implantation of cells producing high amounts of BDNF into the striatum can protect medium spiny neurons and improve motor symptoms in huntingtin mutant mice.

Despite progress in animal studies, the failure rate of drugs in clinical trials to treat Huntington's disease is very high. Fewer than 4 percent have been approved, and they are used to treat symptoms but may not slow disease progression. Drugs commonly prescribed to treat motor symptoms are GABA receptor agonists, such as diazepam, and drugs that reduce dopaminergic neurotransmission, such as tetrabenazine. Drugs prescribed for psychiatric symptoms—anxiety and depression—include serotonin and norepinephrine reuptake inhibitors.

You may be surprised to learn that at the current state of knowledge and technology, it is theoretically possible to eliminate Huntington's disease within the next three decades or so (Mattson 2021). This would occur as follows. All teenage children born to a parent with Huntington's disease would provide a cheek swab or blood sample, which would be used for genetic testing to see if they have inherited the mutant huntingtin gene. If so, they would be counseled on two ways in which they can ensure that they have no children with Huntington's. One way is to choose to have no children and instead consider adoption. But they can also choose to have their own children but still know that they will not have Huntington's disease by means of in vitro fertilization technology. The husband would provide a semen sample, and the wife would provide eggs. Several of the prospective mother's eggs would be fertilized with the father's sperm in a culture dish. The fertilized eggs will divide into two cells. After two more divisions, there will be eight cells in each developing embryo. In each embryo, one of those cells will then be removed and genetically tested to see whether it has a mutant huntingtin gene. An embryo without the mutant huntingtin gene would then be implanted into the mother's uterus. Nine months later the child will be born and will grow up knowing that they will not develop Huntington's disease. Gene mutations that cause early-onset Alzheimer's and Parkinson's diseases could also be eliminated using this approach.

However, elimination of Huntington's or any other dominantly inherited disease would require motivation and cooperation from members of the affected families. It turns out that many people do not want to know whether they have inherited a mutated gene or not. Others may object on religious grounds or may view genetic testing and embryo selection as akin to the Nazis' efforts in eugenics. Using genetic testing for the purpose of preventing a fatal disease is much different from what the Nazis did, however. My opinion is that all children of child-bearing age who have a parent or grandparent with Huntington's disease should be made aware of the fact that they have the ability to choose to have children who will not have Huntington's disease. Should they choose in vitro fertilization and embryo selection, the cost could be covered by insurance.

Promising gene therapy approaches are being developed to either silence the mutant huntingtin gene or to correct the mutation. Perhaps the most exciting prospect for correcting genetic defects in humans involves CRISPR/Cas9 gene-editing technologies (the acronym CRISPR stands for "clustered regularly interspaced short palindromic repeats"). Although the details of this technology are beyond the scope of this book, suffice it to say that the CRISPR/Cas-9 system could enable the excision of the huntingtin mutation and replacement with the nonmutant gene. Such gene therapy technologies could, at least in theory, be applied to any neurological disorder caused by a genetic mutation. Silencing the mutant huntingtin gene in striatal neurons using CRISPR-Cas9 RNA-interference methods have ameliorated striatal neuron degeneration and motor symptoms in mice (H. Liu et al. 2021). Another gene therapy is RNA-interference technology, which uses small 21-base sequences complementary to the targeted gene's messenger RNA sequences. An innocuous virus that produces the desired small interfering RNA and selectively infects neurons, such as adeno-associated virus, can be infused into the brain. This technology, which enables continuous production of the small interfering RNA in neurons, is effective in halting neuronal degeneration in mice with mutant huntingtin (Stanek et al. 2014).

ALS

Lou Gehrig was one of the best players in the history of baseball. He had a batting average of .340 during his 17-year career and was nicknamed the "iron horse" because he never missed a game. From 1923 through 1937, Gehrig's batting average was consistently in the range of .330 to .360, but in 1938 it dropped to .295 and in 1939 plummeted to .143. His muscles began to deteriorate rapidly, and he died in 1941 at the age of 37. The disease Gehrig succumbed to is now called "amyotrophic lateral sclerosis," or ALS. In most instances, a person with ALS will die within three years of diagnosis. However, some people develop a more slowly progressing disease and may survive for decades. Perhaps the best-known example of such a person is the British physicist Stephen Hawking, who was diagnosed with ALS at the age of 21 and died at 76. When Hawking was in his late thirties, he became confined to a wheelchair, and his speech began to deteriorate, although with the help of computer technologies he was able to continue to speak and write.

ALS is a relatively rare but always fatal disease, accounting for approximately one in every 20,000 deaths in the United States. What happens in ALS is that the motor neurons that activate muscle cells degenerate, resulting in muscle weakness and atrophy. Those motor neurons are located in the spinal cord, and their axons extend through peripheral nerves to the muscles. For example, the axons of the motor neurons that innervate muscles in the legs pass through the sciatic nerve. By the time this weakness becomes evident, many motor neurons have already died or are in the process of dying. Motor neurons that innervate muscles in the limbs typically degenerate first, and the person will become unable to walk or lift. The motor neurons that control breathing will eventually degenerate, and the person will die as a result.

As is true for Alzheimer's and Parkinson's diseases, in most instances the cause of ALS is unknown. But approximately 10 percent of cases of ALS are caused by gene mutations (G. Kim et al. 2020). Mutations in the chromosome 9 open reading frame 72 (C9ORF72) gene and transactive response DNA-binding protein 43 (TDP-43) account for most cases of inherited ALS, while mutations in the genes encoding superoxide dismutase 1 (SOD1) and the fusion protein (FUS) account for about 1 percent of cases. SOD1, TDP-43, and FUS mutations are gain-of-function mutations such that mutant proteins have a detrimental action on motor neurons that is unrelated to the normal function of the protein. These mutations result in the accumulation of aggregates of SOD1 and TDP-43 in motor neurons. The C9ORF72 mutations are unusual in that the mutation is a hexanucleotide repeat in a noncoding region of the gene. Presumably, C9ORF72 mutations alter the expression of genes in a manner detrimental for motor neurons.

Early in the disease process, patients with ALS typically experience muscle fasciculations (twitches) and cramps. These symptoms are consistent with hyperexcitability of lower motor neurons. Measurements of the electrical activity of lower motor neurons have consistently demonstrated hyperexcitability prior to development of muscle weakness. This is true both in people with ALS of unknown cause and in those who inherited a SOD1, C9ORF72, or FUS mutation. Other studies have shown that upper motor neurons in the brain exhibit hyperexcitability early in the ALS disease process. Hyperexcitability of these glutamatergic upper motor neurons is likely to contribute to the increased activity of lower motor neurons (figure 8.3). Studies using transcranial magnetic stimulation have also shown that there

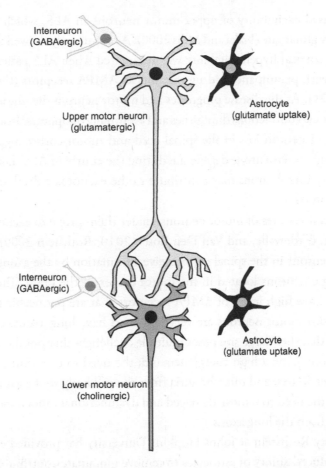

Interneuron
(GABAergic)

Upper motor neuron
(glutamatergic)

Astrocyte
(glutamate uptake)

Interneuron
(GABAergic)

Astrocyte
(glutamate uptake)

Lower motor neuron
(cholinergic)

Figure 8.3
Neurons involved in ALS. Glutamatergic upper motor neurons in the cerebral cortex extend long axons into the spinal cord, where they synapse on the dendrites of cholinergic lower motor neurons. GABAergic interneurons in the motor cortex and spinal cord inhibit upper and lower motor neurons. Astrocytes remove glutamate from synapses.

is increased excitability of upper motor neurons in ALS, which is likely driven by glutamate (Naka and Mills 2000). A recent study showed that such cerebral cortical hyperexcitability was attenuated when ALS patients were treated with perampanel, a drug that blocks AMPA receptors (Oskarsson et al. 2021). As the disease progresses and motor neurons die, the ability to elicit muscle contractions diminishes and eventually disappears. In addition, analyses of neuron loss in the spinal cord and motor cortex suggest that GABAergic interneurons degenerate during the course of ALS. Loss of the inhibitory interneurons may contribute to the excitotoxic death of motor neurons in ALS.

Several features of motor neurons render them prone to excitotoxicity (Bogaert, d'Ydewalle, and Van Den Bosch 2010; Rothstein 2009). Lower motor neurons in the spinal cord receive stimulation by the axons of glutamatergic neurons located in the motor cortex of the brain. The motor neurons have high levels of AMPA receptors that are permeable to Ca^{2+}. Spinal cord motor neurons are very large and have long axons. One can imagine that their large size poses multiple challenges that put them at risk for excitotoxicity: a high energy demand; the need to transport proteins, messenger RNAs, and mitochondria from the cell body to the axon terminal; and the need to remove damaged and dysfunctional mitochondria and proteins from the long axons.

Jeffrey Rothstein at Johns Hopkins University has provided evidence that a reduced ability of astrocytes to remove glutamate contributes to the excitotoxic degeneration of motor neurons in ALS. He and his collaborators found that there is a reduction in the amount of the astrocyte glutamate transporter-1 (GLT-1) in the spinal cords of ALS patients (Rothstein, Van Kammen, et al. 1995). When GLT-1 levels were reduced with antisense oligonucleotides in rat spinal cord slice cultures, motor neurons degenerated (Rothstein, Dykes-Hoberg, et al. 1996). Moreover, experimental reduction of GLT-1 levels in rats resulted in progressive neuronal degeneration and paralysis akin to ALS.

Transgenic mice expressing a SOD1 gene mutation exhibit features remarkably like those of humans with ALS. They develop a rapidly progressing loss of lower motor neurons and muscle weakness, ultimately becoming paralyzed and dying. Their motor neurons accumulate aggregates of SOD1. Evidence suggests that SOD1 mutations result in impaired mitochondrial function in neurons and may also impair astrocytes' ability to remove glutamate from synapses. Drugs that reduce glutamatergic neurotransmission can slow disease progression in SOD1 mutant mice (Tortarolo et al. 2006).

As noted earlier, the causes of most cases of ALS are unknown, but one curious case provides clues (Spencer 2022). Before the end of the twentieth century, many people on the island of Guam suffered and died from a fatal neurodegenerative disorder characterized by impaired control of body movements, progressive paralysis, and dementia. Neurologists dubbed this disorder the "ALS/Parkinsonism Dementia Complex of Guam" (ALS-PD). Early on, it was clear that it was not an inherited disorder and was instead caused by an environmental factor. As Guam has become westernized, the disorder has all but disappeared and does not occur in Guamanians who have moved to the United States. Research suggests that consumption of foods containing the excitotoxin beta-n-methylamino-L-alanine (BMAA) might have been responsible for ALS-PD (Banack, Caller, and Stommel 2010). BMAA is found in low amounts in the seeds of the cycad tree, which are a major source of food for the "flying fox," a species of fruit bat. It turns out that the bats bioaccumulate high amounts of BMAA in their fat and that the Guamanian natives consumed the bats. Analyses of the brains and spinal cords of people who regularly consumed bats and died from ALS-PD revealed the presence of BMAA. Cell culture and animal studies suggest that BMAA can cause excitotoxic damage to motor neurons, but definitive evidence for a causal role in ALS-PD of Guam is lacking.

One drug shown to slow the progression of ALS is riluzole (Miller et al. 1996). This drug acts by enhancing the removal of glutamate from synapses by astrocytes. However, riluzole only slows the disease progression but does not halt it. This is likely because by the time someone is diagnosed with ALS, it is already too late to save many motor neurons. In addition, riluzole only lessens somewhat but does not prevent the activation of glutamate receptors on motor neurons.

Psychiatric disorders are common and often debilitating. These disorders include depression, anxiety disorders, post-traumatic stress disorder, autism spectrum disorders, and schizophrenia. This chapter focuses on the involvement of alterations in the structure and function of glutamatergic neuronal networks in these brain disorders.

ANXIETY, DEPRESSION, AND PTSD

I've suffered through depression and anxiety my entire life; I still suffer with it every single day. I just want these kids to know that the depth that they feel as human beings is normal. We were born that way. This modern thing, where everyone is feeling shallow and less connected? *That's* not human.
—Lady Gaga, quoted in "Lady Gaga: 'I've Suffered through Depression and Anxiety My Entire Life,'" by David Renshaw

Traditionally, many associate PTSD as a condition faced by brave men and women that serve countries all over the world. While this is true, I seek to raise awareness that this mental illness affects all kinds of people, including our youth. . . . No one's invisible pain should go unnoticed.
—Lady Gaga, quoted in "Lady Gaga Developed PTSD after She Was 'Repeatedly' Raped at 19," by Claire Gillespie

Occasional anxiety about certain situations is not unusual and evolved to be important for survival. For example, mice and rats will avoid open spaces in which they are more likely to be easily observed and captured by birds of prey. Some temporary anxiety can be a beneficial motivating factor for people in anticipation of their performances in athletic events, public speaking, or exam taking in school. But excessive and sustained anxiety can be detrimental. Even though I have given many lectures, I still get a little anxious right before each one. But shortly after I begin to talk, that anxiety disappears as I focus on what I am saying. Similarly, when I used to race harness horses or do motocross, I would get nervous in the minutes preceding the start of the race, but once the race began, the anxiety disappeared, and I was completely focused. But I have also experienced periods of more prolonged anxiety revolving around the uncertainty of future events of great importance to me but not under my control. Three times during my life, such chronic anxiety led to bouts of depression that lasted for several months.

Nearly one in every five people in the United States will suffer with an anxiety disorder during their lifetime. About 17 million or one in every 20 people in the United States experienced a bout of major depression in the past year. Only about 40 percent of the people who suffer from an anxiety disorder or depression receive treatment. Nearly one in 10 people in the United States will have PTSD during their life. Women are approximately twice as likely as men to be diagnosed with an anxiety disorder, depression, or PTSD. This gender difference is likely due in part to the fact that men with these disorders are less likely to seek professional help and so go undiagnosed and underreported.

Generalized anxiety and panic attacks are the two most common types of anxiety disorders. Generalized anxiety is characterized by persistent feelings of restlessness, fatigue, difficulty concentrating, muscle tension, excessive worry, and sleep problems. People with such chronic anxiety often have problems in school, work, and social interactions. Panic attacks are sudden periods of intense fear that are typically accompanied by accelerated heart rate, trembling, and feelings of impending doom. About 7 million people in the United States are currently affected by generalized anxiety disorder, and about 6 million have panic attacks.

Everyone experiences periods of sadness that last a few days but do not have a major negative impact on their work, social life, or family. A person is considered to have clinical depression when they have the following types of symptoms every day for a period of several weeks or more: feelings of sadness and hopelessness; irritability often accompanied by angry outbursts; sleep problems; tiredness and weakness; excessive anxiety; difficulty concentrating; feelings of self-blame and worthlessness; recurrent thoughts of death and suicide.

PTSD is perhaps best known as a problem experienced by soldiers who have been in combat, but anyone can develop PTSD. Symptoms usually start within a month of a traumatic event but may not appear until years later. They include recurrent unwanted memories and flashbacks of the traumatic event, upsetting dreams and nightmares, and severe emotional distress. These symptoms are often accompanied by chronic anxiety and depression. People with PTSD will avoid activities, movies, and people or places that remind them of the traumatic event. They will have problems in social and work situations and in relationships. Their ability to perform usual daily tasks is often impaired.

As with most health issues, both environmental factors and genetics influence whether a person develops an anxiety disorder or depression. However, in contrast to studies of many other neurological disorders, such as Alzheimer's, Parkinson's, and Huntington's diseases, geneticists have not identified gene mutations that cause anxiety disorders or depression. Nevertheless, these disorders tend to run in families, and there is some evidence that certain genetic factors influence a person's risk.

Excessive stress is a major factor in anxiety disorders and depression and causes PTSD. The stress may be one severely traumatic event or a more insidious daily stress associated with social interactions, the work environment, or health problems. The COVID-19 pandemic has killed more than 1 million Americans and has left many others with long-term health problems. Moreover the stresses associated with the virus and its consequences—social isolation, loss of income, much uncertainty and loss of control—are profound. It is likely that many people who would have otherwise not developed an anxiety disorder or depression will do so as a result of their COVID experiences.

At this point in *Sculptor and Destroyer*, you may not be surprised to learn that anxiety disorders, depression, and PTSD involve abnormalities in glutamatergic neuronal circuits. In fact, aberrant activity of glutamatergic neurons in certain brain regions results in changes in the structure of neuronal networks and associated symptoms in these disorders. I first consider the current state of knowledge about glutamate and mood disorders and then describe how this knowledge is being translated into more effective treatments.

Neuroscientists have made major progress in understanding the alterations in the brain and body that occur in response to prolonged uncontrolled stress—a risk factor for anxiety disorders, depression, and PTSD. The kinds of stress that lead to a mental disorder are generated in situations from which the individual is unable to remove themselves—physically or mentally. This results in the activation of neuronal networks that evolved to enable adaptive behavioral responses to acute stress, such as an encounter with a predator. Such responses include feelings of fear and anxiety and prompt efforts to end the stress by, for example, running or fighting. These stress-responsive networks include glutamatergic neurons in several different interconnected brain regions, with the hippocampus, hypothalamus, amygdala, and prefrontal cortex being particularly important.

Stressful experiences are often better remembered than nonstressful ones. This makes sense from an evolutionary perspective as the memory of stress facilitates the avoidance of similar negative experiences in the future. But when those bad memories are repeatedly recalled long after the stressful event, mental health can be compromised, and general health adversely affected. As described in chapter 4, the hippocampus is the brain's learning and memory hub. The hippocampus has direct connections to the amygdala and the hypothalamus. Activation of neurons in the hypothalamus during stress results in the release of the hormone ACTH into the blood (figure 4.3). ACTH travels to the adrenal gland, where it stimulates the production of cortisol. Cortisol plays important roles in the body's responses to acute stress by working with adrenaline to increase glucose levels during that period of stress. But chronic activation of this stress-responsive neuroendocrine system can have detrimental effects on the body and brain.

People with generalized anxiety disorder usually have higher cortisol levels compared to people without this problem (Fischer 2021). Studies of rats and mice have shown that corticosterone (the rodent equivalent of cortisol) contributes to the adverse effects of chronic stress on the brain (Popoli et al. 2011). Corticosterone increases the activity of glutamatergic neurons in the hippocampus and renders them vulnerable to dendritic pruning. Evidence suggests that these adverse effects of stress and corticosterone result from increased glutamate release, increased amounts of synaptic AMPA receptors, and impaired glutamate removal by astrocytes.

Brain-imaging studies have revealed that generalized anxiety disorder, PTSD, and depression are associated with a reduction in the size of the hippocampus. However, this finding does not discriminate between two possibilities: that these disorders cause a decrease in the size of the hippocampus or that people with relatively smaller hippocampi are more prone to these disorders. But experiments in animals have shown that the experiencing of chronic stressful situations that cause a heightened state of anxiety or depression causes a decrease in the size of the hippocampus. The decrease in size of the hippocampus apparently results from reductions in the lengths of dendrites and the numbers of synapses on those dendrites. This research was pioneered by the late Bruce McEwen (McEwen, Bowles, et al. 2015). Some of the stresses used in such animal studies are very relevant to humans and include social isolation, chronic sleep deprivation, repeated exposures to cold water, or a randomly applied shock to the feet. Over a period of several weeks or more, such chronic stress causes the dendrites of neurons in the hippocampus to regress and some of their synapses to degenerate.

Elizabeth Gould showed that chronic stress can also cause a decrease in neurogenesis—the production of new neurons from stem cells—in the dentate gyrus of the hippocampus (Gould and Tanapat 1999). The atrophy of neurons and reduced neurogenesis result in a decrease in the size of the hippocampus and an impaired learning and memory ability. Behaviorally, the animals initially exhibit increased anxiety, which then segues into a depression-like state called "learned helplessness." Studies have shown that corticosterone contributes to the atrophy of the hippocampus and associated cognitive impairment caused by chronic uncontrollable stress. For example, Theodore Dumas and coworkers used a gene therapy approach to show that reduction in levels of corticosterone can ameliorate synaptic dysfunction and thus ameliorate spatial learning and memory deficits in rats (Dumas et al. 2010).

The atrophy of neurons that occurs in animal models of anxiety disorders and depression apparently results from a relatively mild type of excitotoxicity in which some synapses are lost but neurons do not die. These adverse effects of chronic stress on the structure and function of neuronal networks are largely reversible in animal models of anxiety and depression. Studies suggest that the adverse effects of chronic anxiety and depression on the hippocampus are also reversible in humans. For example, Yvette Sheline and others measured the size of the hippocampus in patients with depression who had or had not been treated with antidepressants. They found that the size of the hippocampus is significantly smaller in patients with untreated depression compared to those who recovered from depression in response to treatment (Sheline et al. 2012).

Further evidence that hyperactivation of glutamatergic neurons in the hippocampus contributes to anxiety disorders comes from studies of the effects of regular exercise and intermittent fasting on anxiety. Studies in humans have shown that exercise has potent beneficial effects in the prevention and treatment of anxiety disorders and depression (Ashdown-Franks et al. 2020). Animal studies have shown that running-wheel exercise reduces anxiety levels and depression-like behaviors in rats and mice. Elizabeth Gould and colleagues at Princeton University found that running ameliorates the anxiety associated with stressful experiences in mice (Schoenfeld et al. 2013). The reduced anxiety was associated with increased activity of GABAergic inhibitory neurons and a consequent reduction in the activity of glutamatergic neurons in the hippocampus. This effect of exercise on hippocampal neurons explained the reduced anxiety because when the drug bicuculline, which inhibits GABA receptors, was injected into the hippocampus, the anxiety-reducing effect of running was abolished.

Neurons in the hippocampus are reciprocally connected to neurons in the amygdala and prefrontal cortex. A major function of the amygdala concerns behavioral and neuroendocrine responses to stressful situations—it is the brain's "fear center." Studies of animal models and humans suggest that the size of the amygdala may increase in those with anxiety disorders and PTSD (McEwen, Eiland, et al. 2012). In the case of depression, some circuits within the amygdala may atrophy, whereas others may grow. Experiments with animal models revealed that chronic anxiety results in growth of glutamatergic neurons and increased numbers of synapses in some parts of the amygdala. These structural changes are associated with increased behavioral responses to fearful situations and increased aggression between animals living in the same cage. In people with generalized anxiety disorder, functional neuroimaging studies have documented hyperactivation of the amygdala, hypoactivation of the ventromedial prefrontal cortex, and alterations in functional connectivity between these two brain regions. Abnormal neuronal network activities in these brain regions are thought to result in impaired threat detection, emotion control, executive functions, and decision-making.

Geneticists have identified some genes that may affect a person's vulnerability to developing anxiety disorders, depression, and PTSD (Gatt et al. 2015; Lacerda-Pinheiro et al. 2014), including variants in the genes encoding a GABA receptor protein, a Ca^{2+} channel, and BDNF. It is not known exactly how these genetic factors predispose a person to anxiety and depression. However, influences on glutamatergic neurons are likely given that GABA, Ca^{2+} channels, and BDNF all affect the function of glutamatergic neurons.

There is considerable evidence that reduced BDNF production contributes to the neuronal atrophy and impaired neurogenesis that occurs in depression. Postmortem analyses of the brains of people who died suddenly have shown that BDNF levels are lower in the hippocampus of people with depression compared to those without depression (Dwivedi 2010). One study showed that BDNF levels in the cerebrospinal fluid increase after electroconvulsive therapy in patients with depression (Mindt et al. 2020). Another study reported that levels of BDNF in the cerebrospinal fluid are lower in depressed patients with cognitive impairment compared to depressed patients without cognitive impairment (Diniz et al. 2014). Animal studies have shown that chronic stress causes a reduction in BDNF levels in the hippocampus, a reduction that is associated with neuronal atrophy and impaired neurogenesis (Schmidt and Duman 2007). Antidepressant drugs and exercise increase BDNF levels in the hippocampus and stimulate the formation of new synapses and neurogenesis.

GABA, serotonin, and norepinephrine all play important roles in modulating the activities of glutamatergic neurons in neuronal networks affected in anxiety disorders, depression, and PTSD. This is clearly demonstrated by the effectiveness of drugs that affect these neurotransmitters in anxiety disorders, PTSD, and depression. Historically, the first drugs prescribed for patients with anxiety disorders were those that activate GABA receptors, such as diazepam (Valium). Other drugs effective for anxiety disorders include gabapentin and pregabalin, which apparently act by inhibiting presynaptic Ca^{2+} channels, resulting in decreased glutamate release from presynaptic terminals. More recently, drugs initially developed for the treatment of depression have been shown effective in treating anxiety disorders and PTSD. Such antidepressants are called serotonin selective reuptake inhibitors (SSRIs) or serotonin/norepinephrine selective reuptake inhibitors (SNSRIs). SSRIs include fluoxetine and paroxetine, and SNSRIs include duloxetine and venlafaxine. These drugs inhibit the transporter proteins that normally remove serotonin and norepinephrine from synapses and thereby increase the amounts of serotonin and/or norepinephrine at synapses. The serotonin and norepinephrine synapses are located on the dendrites of glutamatergic neurons throughout the brain, including those in the hippocampus, amygdala, and prefrontal cortex.

What about drugs that act on glutamate receptors? Might they be beneficial for people with a mood disorder? In the 1990s, neuroscientists in my laboratory who performed surgeries in rats or mice injected them with an anesthetic called "ketamine." Veterinarians still often use ketamine for anesthesia. However, ketamine is not ideal for surgeries that will last more than 30 minutes or so because it requires multiple doses to keep the animal anesthetized, so inhalation anesthetics such as isoflurane are more commonly used nowadays.

But it turns out that lower doses of ketamine can alleviate depression. The results of the first randomized double-blind controlled trial of ketamine in patients with depression were reported in 2006 (Zarate, Singh, et al. 2006). Their symptoms were evaluated before the treatment and at three and seven days after it. Symptoms improved in the patients treated with ketamine but did not improve in the patients receiving the placebo. Since then, several larger studies have shown that ketamine can rapidly alleviate the symptoms of depression and that this antidepressant effect can last for several weeks or even months after the treatment. Ketamine treatment has also been shown to benefit patients with bipolar disorder (Zarate, Brutsche, et al. 2012). Evidence suggest that ketamine acts on the NMDA glutamate receptor in a manner that reduces opening of the receptor's ion channel. Drugs that affect glutamate neurotransmission may be the future of treatments for anxiety disorders and depression.

SCHIZOPHRENIA

Imagine if you suddenly learned that the people, the places, the moments most important to you were not gone, not dead, but worse, had never been. What kind of hell would that be?
—Dr. Rosen in the film *A Beautiful Mind*

John Nash was a mathematician from West Virginia who made major contributions to game theory, differential geometry, and decision-making in complex systems of relevance to daily life. All of his major academic achievements were made when he was in his early twenties. Soon after joining the faculty at MIT, Nash began experiencing paranoid delusions and was admitted to a psychiatric hospital for the treatment of schizophrenia. He was prescribed antipsychotic drugs, and he had to quit his job. With the help of his wife and his own fortitude, Nash stopped taking the drugs and learned to ignore his delusions. He moved to Princeton University, where he was first allowed to work without having to teach and was later able to teach. In 1994, Nash was awarded the Nobel Prize in Economics in recognition of his contributions to the analysis of equilibria in noncooperative games.

Persons with schizophrenia will have "positive symptoms," including delusions and paranoia. They may believe others are out to get them or that special messages are being sent to them through a radio, TV, or the internet. They also have "negative symptoms," including loss of motivation to hold a job, difficulty showing emotions, and withdrawal from social interactions. They commonly have difficulty concentrating and may also have impaired learning and memory. Obviously, the disease has a major detrimental impact on their lives. Every year, more than 200,000 people in the United States are diagnosed with schizophrenia. The symptoms usually begin when they are between the ages of 18 and 32 and they experience their first episode of psychosis, during which they lose touch with reality and see, hear, or believe things that exist only in their minds.

Studies of the brains of people with schizophrenia point to excessive activation of dopaminergic neurons in the midbrain (Soares and Innis 1999). The dopaminergic neurons that are involved in schizophrenia have synaptic connections with glutamatergic neurons in the hippocampus, prefrontal cortex, and other regions of the cerebral cortex. The first drugs used to treat schizophrenia inhibit dopamine receptors. These drugs, such as haloperidol, chlorpromazine, and resperidone, are effective in reducing hallucinations and delusions but often exacerbate cognitive impairment and emotional blunting.

Increasing evidence, however, suggests that schizophrenia is primarily a glutamate disorder and not a dopamine disorder (Uno and Coyle 2019). Schizophrenia involves widespread dysfunction of glutamatergic neuronal networks throughout the cerebral cortex and limbic system. Examination of the brains of people with schizophrenia and animal models of schizophrenia have revealed alterations in glutamatergic neurons in the prefrontal cortex and hippocampus (Uno and Coyle 2019). Moreover, drugs that activate dopamine receptors do not reproduce the symptoms of schizophrenia. On the other hand, drugs such as ketamine and phencyclidine (PCP) can cause psychosis (Merritt, McGuire, and Egerton 2013). In fact, if a psychiatrist were to evaluate someone who had just taken ketamine or PCP, that doctor would very likely diagnose them as having schizophrenia. In the 1960s, the mechanism of ketamine's and PCP's action was not known. We now know that their effects result from inhibition of NMDA glutamate receptors. Finally, schizophrenia tends to run in families, and geneticists have identified several genes that are associated with this brain disorder. Most of those genes encode proteins that are known to affect glutamatergic neurotransmission (Yuan et al. 2015).

Evidence suggests that the abnormalities in neuronal network excitability that occur in schizophrenia begin during development in the mother's womb or during the early postnatal period (Estes and McAllister 2016). A woman who experiences an infection during pregnancy is more likely to have a child who develops schizophrenia compared to women who have an uncomplicated pregnancy. Animal studies have shown that maternal immune activation during pregnancy and early postnatal social isolation of pups result in schizophrenia-like behaviors that do not emerge until after puberty. These abnormal behaviors include locomotor hyperactivity, increased anxiety, increased responses to novel situations, and cognitive impairment. This is very interesting because in humans with schizophrenia the onset of the symptoms does not occur until after puberty.

Among neuroscientists and pharmacologists, there is currently much interest in the development of new drugs for schizophrenia that target NMDA receptors. Examples of such potential treatments include the amino acids glycine and D-serine, which are known to interact with NMDA receptors. The results of initial clinical trials with glycine and D-serine demonstrated beneficial effects in improving the negative symptoms in people with schizophrenia (Kantrowitz et al. 2015).

AUTISM SPECTRUM DISORDERS

Someone with autism has behavioral problems that typically include social withdrawal and repetitive behaviors. These symptoms are often accompanied by deficits in language development and cognition.

When I was in high school in the early 1970s, I do not recall ever hearing the word *autism* and did not know any children with autism. Indeed, according to the US Centers for Disease Control, in 1980 the prevalence of reported cases of autism was one in 10,000. The term *autism spectrum disorders* (ASDs) was introduced in 2013 to include several related developmental disorders. By 2013, nearly one in 50 children was diagnosed as having an ASD. This dramatic increase is due in part to increased awareness of symptoms in children and to the broadening of the criteria for diagnosis. But research suggests that there has also been a highly significant increase in the actual incidence of ASDs among children during the past 40 years.

What might explain the recent increase in the incidence of ASD? One possibility supported by a rapidly growing body of evidence is that a child is more likely to develop an ASD if they are born to a mother who was metabolically unhealthy during her pregnancy (Rivell and Mattson 2019). A woman with obesity, prediabetes, or diabetes is more likely to give birth to a child with an ASD compared to a woman who has a healthy body weight and is physically fit during her pregnancy. Epidemiological data point to a scenario in which the increased consumption of sugars (in particular fructose) and processed foods, combined with increasingly sedentary lifestyles, has resulted in corresponding increases in maternal obesity and insulin resistance and in turn in increases in ASD (figure 9.1). Animal studies support this scenario. When female rats are fed a diet rich in sugars and saturated fats prior to and during pregnancy, their offspring exhibit behaviors like those of children with an ASD, including social withdrawal and repetitive behaviors.

The evidence for neuronal network hyperexcitability in ASDs is very strong (Frye et al. 2016; Rubenstein and Merzenich 2003). Children with an ASD are significantly more likely to develop epileptic seizures compared to their peers who do not have an ASD. EEG recordings of brain activity reveal abnormal bursts of electrical activity in people with ASD. The EEG

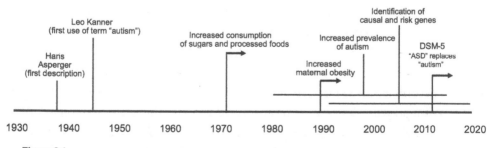

Figure 9.1
Timeline showing important events in the history of research on autism.

abnormalities occur more often in people with ASD who have more severe symptoms and poor intellectual function.

In 1924, the German physiologist Hans Berger, the inventor of the EEG, provided the first evidence that neuronal networks in the brain are active even when a person is resting in a quiet room with their eyes closed. But EEG recordings are made using electrodes on the surface of the brain and provide little information as to where in the brain the electrical signals are emanating. The advent of fMRI enabled us to answer the question of which brain regions are active during rest. Those brain areas are now called the "default mode network," or DMN, and they lie in the medial (middle) regions of the cerebral cortex that form an arc surrounding the corpus callosum. These brain regions include the cingulate gyrus, the precuneus, and the retrosplenial cortex. The activity of glutamatergic neurons in DMN circuits is high when a person is resting quietly with their eyes closed and then decreases when the person interacts with their environment. Neuroscientists believe that the DMN may be the neurobiological basis for "the self" because the DMN seems to function in processing autobiographical information, remembering the past and thinking about the future, thinking about the thoughts of others, making social evaluations, and engaging in moral reasoning.

Imaging of the brains of children and adolescents with an ASD using fMRI have provided evidence for the hyperactivity of some neuronal networks in ASDs. Evidence further suggests that the strength of synaptic connectivity between some brain regions is altered in ASDs, with increased connectivity of neurons in the DMN with neurons in the executive control network—which includes the prefrontal cortex, inferior parietal cortex, and hippocampus—being particularly robust (Abbott et al. 2016). As described in chapter 4, executive functions include reasoning, planning, and problem-solving. Interestingly, the executive control network also includes the frontal eye fields, which are involved in the control of visual attention and eye movements. Children and adults with ASDs very commonly will not make eye contact with other people, which contributes to their social withdrawal.

Neuroscientists have made excellent progress in understanding what is happening in the brain during development in the womb that predisposes a child to an ASD. Their findings are summarized in the following sentence from a recent review article written by me and Aileen Rivell: "Studies of humans and animal models suggest that ASDs involve accelerated growth of neuronal progenitor cells and neurons [during fetal brain development] resulting in aberrant development of neuronal circuits characterized by a relative GABAergic insufficiency and consequent neuronal network hyper-excitability" (Rivell and Mattson 2019, 709).

In ASDs, the growth of the brain occurs more rapidly than is usual. Children who develop an ASD have, on average, bigger brains at birth than children who do not develop an ASD. This greater growth is apparently due to the accelerated production of neurons from stem cells and the accelerated growth of the dendrites and axons of neurons. The formation of synapses between glutamatergic neurons as well as between GABAergic and glutamatergic neurons is apparently altered in ways that cause an excitatory imbalance.

The cause of accelerated growth of neurons in ASDs may be due to excessive activation of the mechanistic target of rapamycin (mTOR) pathway, a cellular system that controls the production of new proteins from amino acids (Winden, Ebrahimi-Fakhari, and Shahin 2018). Activity of the mTOR pathway is increased by food intake and is reduced by fasting and exercise. When mTOR is active, cells are in a growth mode. When mTOR is inactive, cells are in a "conserve-resources" mode.

Animals, including humans, evolved in environments where cells throughout their body and brain were intermittently switching back and forth between growth and conserve-energy modes. They were hunter-gatherers who had to expend physical energy every day as they worked to find food. They did not have the luxury of well-stocked pantries or refrigerators. And, of course, they included pregnant females. This is to say that during evolution the development of the human brain in utero occurred in mothers who ate intermittently and were physically active. Presumably, the rate of growth of neurons and formation of their synaptic connections was adapted to those maternal lifestyles. It is not known whether ASDs existed in children in hunter-gatherer societies, but I expect it was very rare.

Studies of transgenic mice with gene mutations that cause ASDs in humans have solidified the theory that the behavioral manifestations of ASDs result from neuronal network hyperexcitability (Rivell and Mattson 2019). Two ASDs that have a genetic cause are fragile X syndrome and Rett syndrome. The gene encoding the fragile X mental retardation protein (FMRP) is located on the X chromosome and is mutated in fragile X syndrome. The FMRP mutations are "loss-of-function" mutations. Mice in which the gene encoding FMRP is rendered dysfunctional exhibit behaviors like those of children with fragile X syndrome, including social withdrawal and repetitive behaviors. These mice have increased activity of glutamatergic neurons in the hippocampus. The fragile X mice also have hyperactivation of the mTOR pathway, resulting in overexuberant growth of neuronal dendrites and synapses. Inhibition of the mTOR pathway can reverse the behavioral abnormalities in fragile X mice, suggesting that excessive growth and connectivity of neurons are responsible for the symptoms of ASD (Bhattacharya et al. 2012).

Rett syndrome is an ASD caused by mutations in the gene encoding the methyl CpG binding protein 2 (MECP2, with CpG denoting a DNA sequence) located on the X chromosome. During their first year of life, children with Rett syndrome avoid making eye contact with their parents or others. Between the ages of one and four years, they exhibit impaired acquisition of motor skills and spoken language. They typically have repetitive hand movements to the mouth and random touching and grasping of objects. After the age of two years, additional problems become apparent and include epileptic seizures and social withdrawal. MECP2 functions as a regulator of gene expression, and mice with mutant MECP2 have altered expression of many genes in their brain cells. Studies of such Rett mice have provided evidence for hyperexcitability of glutamatergic neurons in ASDs (W. Li, Xu, and Pozzo-Miller 2016).

Mice with mutations in several different genes involved in glutamatergic or GABAergic neurotransmission exhibit social withdrawal and repetitive behaviors as in ASDs. One such gene encodes the SH3 and multiple ankyrin repeat domains 3 (SHANK3) protein. SHANK3 is located just under the membrane of dendritic spines, where it functions in the coupling of glutamate receptor activation to Ca^{2+} influx. Mutations in SHANK3 cause ASD-like behaviors associated with hyperexcitability of glutamatergic neurons in the hippocampus (Monteiro and Feng 2017). Mice in which the gene encoding a protein critical for the production of GABA is mutated also exhibit ASD-like behaviors (Sandhu et al. 2014). These findings show that shifting the balance between glutamate and GABA toward the glutamate side is sufficient to cause the symptoms of ASDs (Braat and Kooy 2015).

An unusual feature of gene mutations that cause ASDs is that they most commonly occur de novo, while the child is developing in the womb. This fact is by itself compelling evidence that the environment of the developing embryo may determine whether a child is born with an ASD or not. Most mutations result from damage to the DNA caused by free radicals. It therefore seems likely that during the development of neurons in brains predisposed to ASD, there is increased production of free radicals in neurons, reduced antioxidant defenses, impaired repair of the damaged DNA, or a combination of these three irregularities. However, in most instances ASD is not caused by a gene mutation. The question then becomes: How does the environment of the developing embryo influence the likelihood of a baby being born with an ASD in the absence of a gene mutation?

In 1809, the French zoologist Jean-Baptiste Lamarck proposed a theory of inheritance of acquired characteristics that posited that some physical and behavioral features of a person are determined by the behaviors and environment of their parents. For example, a child may suffer from excessive anxiety because their mother experienced a traumatic event and thereafter had chronic anxiety. Or the son of a farmworker will have bigger muscles than the son of an accountant even though both boys engage in the same amount of physical activity. Some other scientists of Lamarck's time had a similar idea, but the theory was soon buried in the literature and ignored as geneticists established that most characteristics are passed in a reliable manner from parents to their children independently of the parents' environment. The discovery of the structure of DNA and the molecular nature of inheritance and mutations seemed to have completely obliterated the possibility of Lamarckian inheritance.

During the past three decades, however, data have accumulated that support Lamarck's hypothesis. One such trait that has been studied intensely is obesity. Epidemiological studies have shown that classical genetic inheritance accounts for only a miniscule number of instances of childhood obesity. Children's diet and amount of physical activity definitely influence whether they develop obesity or not. But children born to a parent who became obese because of a sedentary lifestyle that includes overeating are more likely to develop obesity even when they have a healthy diet and are physically active. When female rats are made obese by feeding them a diet rich in sugar and saturated fats prior to and during pregnancy, their offspring are prone to obesity even when consuming a healthy diet. When female rats are subjected to stress during their pregnancy, their offspring exhibit anxiety and depression-like behaviors, impaired learning and memory, and social withdrawal. These kinds of animal studies have solidified the case for Lamarckian inheritance and have identified specific molecular mechanisms responsible for such inheritance. Those mechanisms occur beyond the genes and are therefore called "epigenetic" mechanisms.

Epigenetic mechanisms result in enduring changes in gene expression that occur without changes to the DNA base sequence of the gene itself. They result from molecular modifications that affect the expression of the gene. An increasing number of such epigenetic mechanisms are being discovered, but the most intensively studied is DNA methylation. The addition or removal of the methyl group CH_3, an alkyl derived from methane, to or from DNA can determine if and to what extent the gene is turned on. In general, high amounts of DNA methylation suppress gene expression. Several studies examined the effects of maternal obesity and stress on DNA methylation in the brains of a developing fetus as well as in the brains of the same mother's adolescent and adult children (Rivell and Mattson 2019). Maternal obesity results in changes in the methylation of genes involved in the regulation of food intake and circadian rhythms. Maternal stress in rats can cause changes in the expression of genes encoding glutamate receptor proteins in their offspring's hippocampus and prefrontal cortex (Verhaeghe et al. 2021). These alterations in glutamate receptors were more pronounced in male offspring, which is interesting because human boys are more likely than girls to develop an ASD.

The evidence that maternal obesity and stress predispose offspring to ASD is important because it provides the opportunity for the counseling of prospective mothers by their physicians and for a broader dissemination of this information to the general public. The general conclusion is that pregnant women should maintain a healthy lifestyle that includes exercise and moderation in energy intake.

10 ONE PILL MAKES YOU . . .

Naturally occurring chemicals that affect mental states—perception, mood, cognition, belief, pain levels, desire, behavioral inhibitions, vigilance, and so on—have been used in religious ceremonies, popular culture, and medicine for millennia. Psychoactive drugs used in religious ceremonies include psychedelics, such as those found in "magic mushrooms." Alcohol, marijuana, and opioids are commonly used in popular culture for the purposes of reducing anxiety and pain, increasing sociality, and feeling happy. Psychostimulants such as caffeine, nicotine in tobacco, and cocaine are used to increase alertness and "energy" levels. Unfortunately, with the exception of psychedelics and caffeine, psychoactive drugs are often abused, resulting in detrimental effects on health and welfare.

According to US Centers for Disease Control data, the death rate from drug overdoses in 1980 was less than two in 100,000. Beginning around 2000, however, an exponential increase in overdose deaths occurred, such that in 2019 the death rate was more than 20 in 100,000, with a vast majority of those deaths caused by opioids. In many cases, the deceased person initially became addicted to an opioid prescribed by a doctor to treat pain. Unfortunately, once a person has become dependent on alcohol, opioids, cocaine, or methamphetamine, it is very difficult for them to recover without relapsing.

Rats and mice can become addicted to the same drugs that destroy the lives of many people. The pleasure-seeking behaviors of laboratory rats and mice can be evaluated in several ways. One commonly used method is to place the animal in a chamber with a drug-delivery tube affixed to the wall of the chamber, from which the animal can sip the liquid containing the drug. On the opposite wall of the chamber is a lever that the animal can press with its paw or nose. The animal will learn that when it presses the lever, more liquid will be in the tube. The animal is placed in the chamber for several hours daily. The number of lever presses in animals receiving the drug in their drinking water is compared to the number of lever presses of a control group of animals that does not receive the drug. Addiction to the drug is indicated by increased lever pressing over a period of several days. If the drug-receiving animals are then no longer provided with the drug, they will exhibit "withdrawal behaviors," such as body shakes, paw tremors, and teeth chattering. During the withdrawal period, both rats and humans exhibit increased activity of glutamatergic circuits in several brain regions, which decreases as withdrawal symptoms subside.

The activities of ventral tegmental dopaminergic neurons and neurons in the nucleus accumbens are increased in addiction. The activity in glutamatergic neuronal circuits in the prefrontal cortex are also increased in association with the urge to partake of the substance to which the person is addicted. This has been shown in people addicted to cocaine, nicotine, gambling, and even highly palatable foods. Rita Goldstein and Nora Volkow at the US National Institute on Drug Abuse believe that the loss of control that occurs in addiction results from aberrant neuronal circuit activity in the prefrontal cortex. In a review article, they came to the following conclusion: "Disruption of the prefrontal cortex in addiction underlies not only compulsive drug taking but also accounts for the disadvantageous behaviours that are associated with addiction and the erosion of free will" (Goldstein and Volkow 2011, 652).

This chapter focuses on the general notion that psychoactive drugs used for recreational or medical purposes exert their effects on mental states and behaviors by altering the activities of glutamatergic neuronal circuits. In most instances, the molecular mechanisms of psychoactive drugs' actions are known, but exactly how they affect thoughts and behaviors at the neuronal network level has not been established. Figure 10.1 illustrates the sites of action of many of the most widely used psychoactive drugs on glutamatergic neurons, and figure 10.2 shows their chemical structures. Ketamine and phencyclidine act directly on NMDA receptors. The

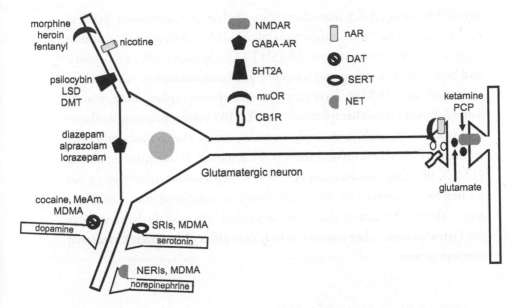

Figure 10.1

Sites of action of psychoactive chemicals used for recreational or medical purposes. 5HT2A = serotonin receptor 2A; CB1R = cannabinoid receptor 1; DAT = dopamine transporter; DMT = N, N-dimethyltryptamine; GABA-AR = GABA receptor A; LSD = lysergic acid diethylamide; MDMA = 3, 4-methylenedioxymethamphetamine; MeAm = methamphetamine; muOR = mu opioid receptor; nAR = nicotinic acetylcholine receptor; NERIs = norepinephrine reuptake inhibitors; NET = norepinephrine transporter; NMDAR = NMDA receptor; PCP = phencyclidine; SERT = serotonin transporter; SRIs = serotonin reuptake inhibitors.

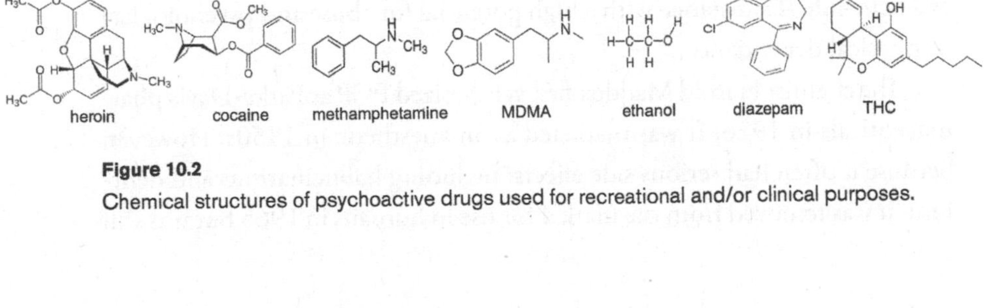

Figure 10.2

Chemical structures of psychoactive drugs used for recreational and/or clinical purposes.

psychedelic drugs 3,4,5-trimethoxyphenethylamine (mescaline), psilocybin, LSD, and N, N-dimethyltryptamine (DMT) activate the serotonin receptor 5-hydroxytryptamine 2A (5HT2A). Opioids such as morphine and heroin activate mu opioid receptors, and benzodiazepines such as diazepam activate GABA-A receptors. Cocaine, methamphetamine, and 3, 4-methylenedioxymethamphetamine (MDMA) inhibit presynaptic dopamine, serotonin, and norepinephrine transporters, thereby increasing the amounts of these neurotransmitters at the synapses. Tetrahydrocannabinol (THC), the main psychoactive chemical in marijuana, activates receptors located on the presynaptic terminals, resulting in reduced release of glutamate. The mechanism of the action of alcohol (ethanol) at the molecular level remains elusive, but it most certainly does affect glutamatergic neuronal network activity.

ANGEL DUST AND SPECIAL K

In popular culture, "angel dust" is a nickname for the drug phencyclidine, or PCP. It is taken by intranasal inhalation or intravenously. The behavioral effects of PCP vary depending on the dose. Low doses cause intoxication like that of alcohol, including unsteady gait, loss of balance, and slurred speech. Somewhat higher doses produce analgesia and sedation, while high doses can cause convulsions. The psychological effects of PCP include depersonalization, changes in body image, euphoria, hallucinations, and paranoia. A small percentage of PCP users experience psychotic episodes, and some develop schizophrenia after such episodes. There are anecdotal reports of people committing violent acts while under the influence of PCP, but there is no evidence for PCP actually causing aggression in people not otherwise disposed to such behaviors. The US Drug Enforcement Agency lists PCP as a Schedule II substance with a high potential for abuse and psychological or physical dependence.

The chemist Harold Maddox first synthesized PCP at Parke-Davis pharmaceuticals in 1926. It was marketed as an anesthetic in 1950s. However, because it often had serious side effects, including hallucinations and delirium, it was removed from the market for use in humans in 1965 but was still used by veterinarians. It would be more than 20 years before the mechanism by which PCP elicits its effects on brain function was discovered.

In 1988, when I was working as a postdoctoral fellow with Ben Kater at Colorado State University, Michael Bennett, a neuroscientist friend of Ben's, visited the laboratory. At that time, reports of the existence of glutamate receptors that were selectively activated by NMDA had just appeared in the literature. David Lodge had provided evidence that PCP exerted its effects by altering the activation of NMDA receptors (see Lodge et al. 2019). Bennett wanted to confirm those findings and elucidate the underlying mechanism, and Ben asked me to help him. Bennett had made a fluorescent form of PCP that enabled its visualization under a microscope that uses ultraviolet light. We found that when we added the fluorescent PCP to the solution bathing cultured hippocampal neurons, it became associated with the cell membrane. Additional experiments revealed that PCP enters the NMDA receptor channel when the channel opens in response to glutamate binding and membrane depolarization and that PCP reduces Ca^{2+} influx through NMDA receptors (Kushner, Bennett, and Zukin 1993).

Jaakko Paasonen and his colleagues performed fMRI analyses of rats' brains before and after administration of PCP. Their data show that PCP can cause widespread alterations in the functional connectivity of multiple brain regions and that these alterations in brain glutamatergic neuronal network activities were associated with impaired cognition and reduced social interactions (Paasonen et al. 2017).

Recreational drug users often refer to ketamine as "special K" or "vitamin K." In the 1970s, several publications described the psychological experiences of ketamine users, including the comic strip *The Fabulous Furry Freak Brothers* and books by the neuroscientist John Lilly and the writer Marcia Moore. The experiences desired by recreational ketamine users are a sense of detachment from one's body and the external world as well as hallucinations. Ketamine is often the drug of choice for use by people at "raves," or dance parties at a private property or public space with live or electronic music. But ketamine has also been used for nefarious purposes such as date rape. The US Drug Enforcement Agency lists ketamine as a Schedule III drug with less potential for abuse and addiction than Schedule I and II substances.

Functional brain-imaging studies have shown that ketamine causes a rapid increase in functional connectivity between the dorsolateral prefrontal cortex and hippocampus in rhesus monkeys and humans (Maltbie, Kaundinya, and Howell 2017). Chapter 9 described how ketamine can elicit rapid and sustained antidepressant effects. It is unclear whether alterations in the activities of the same neuronal networks are involved in ketamine's psychotomimetic (schizophrenia-like) and antidepressant effects. However, accumulating data suggest that increases in prefrontal cortex connectivity underlie these rapid but transient psychotomimetic effects while more delayed changes in synaptic connectivity may underlie ketamine's sustained antidepressant effect.

Since ketamine and PCP inhibit NMDA receptors, why don't these drugs cause a widespread reduction in neuronal network activity throughout the brain? The answer may be that they inhibit not only NMDA receptors on glutamatergic neurons but also NMDA receptors on GABAergic interneurons. NMDA receptors are present in high amounts in fast-spiking interneurons that function to constrain the activity of glutamatergic neuronal circuits. By inhibiting NMDA receptors on these GABAergic interneurons, PCP and ketamine lessen inhibition of glutamatergic neurons. This mechanism is consistent with evidence that PCP and ketamine can cause psychosis because psychosis in schizophrenia is associated with a deficit in GABAergic inhibition.

Interestingly, evidence is emerging that suggests a potential benefit of ketamine for treating addiction to alcohol and other drugs. Several studies have showed that a single intravenous infusion of ketamine increases abstinence rates in people dependent on alcohol or opioids (Witkin et al. 2020; Worrell and Gould 2021). Whether ketamine will become a common treatment for drug addictions remains to be determined.

PSYCHEDELICS

Psychedelics are a class of hallucinogenic chemicals that induce altered states of consciousness. Humans have used them for thousands of years, most commonly in religious rituals and for spiritual experiences. For example, during ceremonies organized for a specific purpose, such as praying for someone who is ill, Navajos and Indigenous Mexicans consume the crushed tops of the peyote cactus, *Lophophora williamsii*, which contains the hallucinogenic chemical mescaline. In the Amazon forests of western Brazil, Indigenous groups make a preparation called *ayahuasca* by boiling the stems and leaves of the shrub *Psychotria viridis*, which contains DMT, with the vines of *Banisteriopsis caapi*, which contains high amounts of chemicals that inhibit the enzyme monoamine oxidase A (MAOA). When ingested alone, boiled *Psychotria* leaves have negligible psychedelic effects because MAOA in the intestines and liver rapidly degrade DMT, but the *Banisteriopsis* in *ayahuasca* inhibits that degradation.

Drawings found in Spain that date to approximately 6,000 years ago show several mushrooms that mycologists believe are *Psilocybe hispanica*, which contain the hallucinogens psilocin and psilocybin. Mushrooms carved from stone dating to at least 3,000 years ago were found in the ruins of Mayan civilizations in Central America. Numerous ancient Mexican Aztec artworks depict hallucinogenic mushrooms. Many different Spanish explorers documented the use of such "magic mushrooms" in Native ceremonies in the 1500s. Indigenous people in Mexico and Central America continue to use psilocybin-containing mushrooms for religious and ceremonial purposes. More than 200 species of mushrooms produce these hallucinogens.

People who have taken psychedelics often find it difficult to describe the experience. The Harvard psychiatrist John Halpern has participated in ceremonies with peyote as part of his research on psychedelics and says that it instills awe and reverence and reduces anxiety. Several books have been written around experiences with psychedelics. Perhaps the most famous is British writer Aldous Huxley's *The Doors of Perception*, published in 1954. In his book, Huxley describes his experiences while under the influence of mescaline in terms of the distortion or dissolution of spatial relationships and an emphasis on the intensity and significance of visual hallucinations.

But there is considerable variability in each individual's subjective experiences of psychedelics, which can cause profound modifications in mood (joy, euphoria, anxiety, fear), space–time relations, sensory illusions, dissolution of boundaries between oneself and others, and confusion. A person's experience depends on the dose of the psychedelic, with low to moderate doses often resulting in mystical experiences and positive mood and high doses causing fear, anxiety, and paranoid delusions. The environment in which the psychedelics are taken also influences the person's experience, with quiet, peaceful settings fostering positive experiences.

Methods have been developed for synthesizing psychedelic chemicals. The Swiss chemist Albert Hofmann was the first to isolate psilocybin and psilocin from mushrooms and synthesize them from the amino acid tryptophan. Ernst Spath first synthesized mescaline in 1918, and Richard Manske was the first to synthesize DMT. Hofmann first synthesized LSD in 1938. While synthesizing a larger batch in 1943, he accidently absorbed some of the LSD through his skin. He described its hallucinogenic effects as a dream-like state with a kaleidoscopic display of colors and shapes (Hofmann 1980). Interestingly, he is reported to have used small doses of LSD throughout his 102-year-long life, which attests to the fact that this drug does not adversely affect general health (Roberts 2017). Because LSD is very potent and easy to synthesize in large amounts, it became readily available and was widely used by young adults in the US counterculture during the 1960s. Then in 1970, as a result of the pressure from parents and lawmakers, LSD became illegal to manufacture, sell, or use in the United States. At that time, psilocybin had recently become similarly controlled by the federal government. DMT and mescaline are also illegal to possess (except for religious or scientific use), sell, or manufacture.

The illegalization of psychedelic chemicals put a damper on the considerable scientific research being done with them in the 1950s and 1960s. That research was demonstrating psychedelics' beneficial effects for the treatment of anxiety, depression, and alcohol addiction. The first trial of LSD in people with depression was in 1957. This and subsequent trials demonstrated antidepressant effects of both LSD and psilocybin.

The first studies of psilocybin appeared in the scientific literature in 1958. Those studies included a description of psilocybin isolation from mushrooms and its chemical synthesis by Albert Hofmann (Hofmann et al. 1958). Studies describing the psychological effects of psilocybin soon followed. Dose–response studies demonstrated that, as with LSD, psilocybin is very safe, with no toxic effects even at high doses. In the 1960s, some psychiatrists used LSD and psilocybin in conjunction with psychotherapy.

Frederick Barrett, Roland Griffiths, and their colleagues at Johns Hopkins University have been at the forefront of recent clinical research studies of psilocybin. They published the results of two clinical trials of psilocybin in patients with major depression (Davis et al. 2021; Griffiths et al. 2016). In both studies, this psychedelic was effective in improving the patients' mood. The antidepressant effects of psilocybin were realized with only one or two treatments and lasted for at least one month. In another study of young healthy volunteers, brain neuroimaging and behavioral assessments were made before as well as one and four weeks after a single dose of psilocybin. In the authors' words, the results showed that "one-week post-psilocybin, negative affect and amygdala response to facial affect stimuli were reduced, whereas positive affect and dorsal lateral prefrontal and medial orbitofrontal cortex responses to emotionally-conflicting stimuli were increased. One-month post-psilocybin, negative affective and amygdala response to facial affect stimuli returned to baseline levels while positive affect remained elevated, and trait anxiety was reduced" (Barrett et al. 2020). These results suggest that psilocybin can increase positive emotions and decrease negative emotions by modifying the activities of glutamatergic neuronal networks in the prefrontal cortex and amygdala.

How do psychedelics affect neuronal network activity? Which networks are responsible for these chemicals' effects on one's perception? What is the explanation for the distortion or escape from reality experienced during a psychedelic trip? Neuroscientists have made progress toward answering the first two questions, but the answer to the third question remains elusive. What has become clear is that psychedelics exert their hallucinogenic and therapeutic effects by modifying the activity of glutamatergic neurons, in particular those in the prefrontal cortex (Nichols 2016).

What psychedelics have in common is that they activate serotonin 5HT2A receptors located on the dendrites of glutamatergic neurons. Activation of the 5HT2A receptors results in a rapid increase in the activity of the glutamatergic neurons (Mason et al. 2020). This effect of psychedelics can be blocked by drugs that inhibit 5HT2A receptors or glutamate receptors

(AMPA and NMDA receptors). But only certain glutamatergic neurons may exhibit robust responses to psychedelics. Indeed, evidence suggests that only about 5 percent of the glutamatergic pyramidal neurons in the prefrontal cortex respond directly to psychedelics. Studies by David Martin and Charles Nichols at Louisiana State University suggest that the neurons that respond to the psychedelics have up to a tenfold greater number of 5HT2A receptors compared to unresponsive neurons (Martin and Nichols 2016).

Human neuroimaging studies have shown that psychedelics alter the activity of prefrontal cortex neuronal networks that receive sensory input relayed from neurons in the thalamus and sensory cortex (Castelhano et al. 2021). Several studies have shown that the experiences of people taking psychedelics are associated with increased activity of circuits in the frontal cortex and medial temporal lobes. However, psychedelics may also decrease activity in cortical neuronal networks involved in the filtering of sensory information. Normally, the brain attends to features of sensory information that are important for cognition and decision-making. By reducing the prefrontal cortex's filtering function, psychedelics may cause free-floating cascades of images and sounds. The results of neuroimaging studies also suggest that psychedelics alter the activity of neurons in the amygdala, which may explain their ability to reduce anxiety.

Perhaps the most consistent finding from fMRI studies with psychedelics is a decrease in activity in neuronal circuits of the DMN (Nichols 2016). Chapter 9 pointed out that the DMN is thought to be involved in one's concept of being independent from others—"the self." Studies have shown that soon after psychedelics are ingested, there occurs reduced functional connectivity between the DMN and other brain regions, including the hippocampus and amygdala.

The profound perceptual effects of psilocybin, mescaline, and DMT occur within minutes of ingesting them and subside within a few hours. But these psychedelics often have beneficial effects on mood that last for many weeks or longer. The increased activity in glutamatergic neurons in response to psychedelics likely results in increased production of BDNF (de Vos, Mason, and Kuypers 2021). It is not known whether BDNF is involved in the immediate and transient psychedelic experience, but it is very likely involved in the long-lasting antidepressant and antianxiety effects of psychedelics. Chapters 3, 4, and 5 described how activation of glutamate receptors increases the production of BDNF and how BDNF stimulates the growth of dendrites and the formation of new synapses. Evidence from animal studies supports the notion that psychedelics can cause long-lasting changes in the structure of neuronal networks.

Calvin Ly, David Olson, and their colleagues at the University of California, Davis, showed that LSD, DMT, and psilocin stimulate the growth of glutamatergic neurons' dendrites in rat cerebral cortical cell cultures (Ly et al. 2018). Each psychedelic also increased the formation of synapses on the glutamatergic neurons' dendrites. The increased number of synapses was associated with neurons' increased spontaneous electrical activity and increased BDNF levels. A chemical that blocks the BDNF receptor and a chemical that inhibits the mTOR pathway abolished the psychedelics' ability to enhance dendrite outgrowth and synapse formation. Ling-Xiao Shao, Alex Kwan, and their colleagues at Yale University showed that a single dose of psilocybin stimulates the formation of new synapses on dendrites of glutamatergic pyramidal neurons in the frontal cortex of mice (Shao et al. 2021). The new synapses formed within 24 hours of administration of psilocybin, and many of those synapses persisted for at least one month. The formation of new synapses occurred specifically on glutamatergic pyramidal neurons in layer V of the cerebral cortex. The increased number of synapses was associated with an antidepressant effect in a stress-induced learned-helplessness model of depression.

It is interesting that glutamatergic pyramidal neurons in layer V of the cerebral cortex are implicated in psychedelics' psychological effects, specifically their beneficial effects on anxiety and depression. Studies have shown that layer V pyramidal neurons have considerably higher amounts of the serotonin 5HT2A receptor than other neurons in the cerebral cortex, which alone suggests important roles for these pyramidal neurons in the actions of psilocybin, LSD, and DMT. Chronic social stress—a risk factor for anxiety disorders and depression—has been shown to cause the pruning of synapses on dendrites of layer V pyramidal neurons in the prefrontal cortex. This effect of chronic stress is associated with reduced activation of these glutamatergic neurons and with impaired emotional responses and executive functions. Layer V pyramidal neurons provide the major outputs of the prefrontal cortex to the hippocampus, amygdala, and other brain regions. Brain-imaging studies such as those done by Roland Griffiths and colleagues suggest that psychedelics enhance the prefrontal cortex's functional connectivity with other brain regions (Griffiths et al. 2016). Altogether, the available data suggest that activation of 5HT2A receptors on glutamatergic pyramidal neurons in the prefrontal cortex promotes the maintenance and formation of synapses, thereby enhancing one's ability to cope with and overcome stress. This may explain why an increasing number of clinical studies are showing the enduring therapeutic effects of psychedelics in people with anxiety, depression, and PTSD.

PAIN AND OPIOIDS

Opioids are a class of drugs that include morphine and man-made chemicals derived from morphine, such as heroin. Morphine is produced by the opium poppy, *Papaver somniferum*, where it is concentrated in the seed capsules. In 1804, the German pharmacist Friedrich Serturner isolated a chemical from opium poppies and found that it causes sleepiness (Devereaux, Mercer, and Cunningham 2018). He originally named the substance "morphium" after the Greek god of dreams. Pain reduction and euphoria were later shown to be prominent effects of morphine. Only later would it become clear that many people can easily become addicted to opioids such as morphine.

One in every 100 people in the United States has an opioid-use disorder, meaning they have a physical and psychological reliance on opioids. Many of these people have difficulties in maintaining employment. In 2019, there were nearly 50,000 opioid overdose deaths in the United States, most of which involved the synthetic opioids heroin and fentanyl.

Before I elaborate on how opioid dependency occurs, a brief summary of the involvement of glutamatergic neurons in pain perception is in order. In 1983, the British veterinarian and neuroscientist David Lodge was studying the effects of ketamine on dorsal root ganglion neurons' responses to glutamate and NMDA. His research was aimed at understanding how anesthetics reduce pain. He was studying the dorsal root ganglion because that is where the cell bodies of the sensory neurons that convey pain reside. Lodge found that ketamine reduces the neurons' responses to NMDA and glutamate but has little effect on their responses to kainic acid or GABA (Lodge et al. 2019). Since then, neuroscientists established that ketamine gets into the NMDA receptor channel, where it reduces Ca^{2+} influx.

Sensory neurons innervate every tissue in the body. When a tissue is damaged, pinched, or exposed to something very hot, dorsal root ganglion neurons that innervate the tissue are depolarized, and action potentials are triggered. When you prick your finger with a pin or touch a hot stove, your conscious awareness of the pain occurs within a fraction of a second. The pain pathway consists of glutamatergic neurons located in the spinal cord and brain. The glutamatergic sensory neurons in the dorsal root ganglion synapse upon glutamatergic neurons in the spinal cord. The spinal cord neurons send their axons up to the brain, where they synapse upon glutamatergic neurons in the thalamus. The thalamus is a relay station for information entering the brain from all of our senses except smell. Glutamatergic neurons in the thalamus extend their axons into the sensory areas of the cerebral cortex, where the information is encoded and integrated with other relevant information. Pain pathways also involve other regions of the brain, including those concerned with emotions, cognition, and decision-making.

Memories of painful stimuli are important for preventing injury and death. For example, in the case of pain caused by touching a hot stove, the visual representation of the stove becomes strongly associated with the pain in your memory. This is obviously important so that you know to avoid touching the stove in the future. Unfortunately, pain signals can become persistent, as occurs in conditions of chronic tissue injury and inflammation or in some neuropathies. Evidence suggests that chronic pain involves changes in glutamatergic neuronal connections in the spinal cord and brain that result in a persistent activation of neuronal circuits that perceive pain. A memory of the pain persists even in the absence of continuing tissue damage. For example, studies in animal models of chronic pain have demonstrated LTP at pain-conveying synapses in the dorsal horn of the spinal cord.

The anterior cingulate cortex seems to play a particularly important role in chronic pain (Bliss et al. 2016). Neurons in this brain region receive inputs from pain-responsive glutamatergic neurons in the thalamus. This was demonstrated in studies in which electrical stimulation of the thalamus elicited responses in neurons in the anterior cingulate cortex. Experiments in animals have shown that persistent stimulation of glutamatergic neurons in the anterior cingulate cortex results in LTP of synapses that lasts for at least several hours. This chronic-pain-related increase in synaptic strength involves NMDA receptor activation, Ca^{2+} influx, and activation of the transcription factor CREB. The importance of glutamatergic circuits in pain is evident from the clinical efficacy of drugs that reduce glutamatergic neurotransmission. For example, ketamine is very effective in ameliorating both acute and chronic pain.

Pain-related information is also communicated from the amygdala to the anterior cingulate cortex. In rats, stimulation of neurons in the anterior cingulate cortex elicits anxiety-like behaviors. Neuroimaging studies have shown that people with anxiety disorders have increased activity in neuronal networks in the anterior cingulate cortex (Fredrikson and Faria 2013). These interactions between circuits in the amygdala and cingulate cortex likely explain, at least in part, the well-known relationship between pain and anxiety. Pain increases anxiety, and vice versa. Indeed, drugs initially developed for the treatment of anxiety or depression reduce pain in many patients. They include gabapentin and serotonin/norepinephrine reuptake inhibitors such as duloxetine and venlafaxine.

How do opioids reduce pain perception, and why are they so addictive? The answers to these questions revolve around the mu opioid receptor on glutamatergic neurons in several different brain regions (Chartoff and Connery 2014; Corder et al. 2018). Four opioid receptors have been identified—mu, delta, kappa, and nociceptin. These receptors evolved to respond to the endogenous neuropeptides beta-endorphins, enkephalins, and dynorphins. Research has established that among the different opioid receptors, the mu opioid receptor is responsible for both the analgesic and addictive effects of opioids. Activation of the mu opioid receptor results in a rapid (minutes) reduction in cyclic AMP levels and reduced Ca^{2+} influx through voltage-dependent channels. In addition, opioids can activate K^+ channels. In these ways, opioids reduce the activity of glutamatergic neurons in pain pathways.

The mu opioid receptor is present in all glutamatergic neurons in pain pathways—dorsal root ganglion sensory neuron; spinal cord neurons that relay pain messages to the thalamus; thalamic neurons that relay to the somatosensory cortex; neurons in the prefrontal cortex, hippocampus, and amygdala. However, studies in mice have shown that when the mu opioid receptor is selectively deleted from dorsal root ganglion neurons, the animals still respond to painful stimuli. Therefore, the analgesic effect of opioids is apparently mediated by neurons upstream of the nociceptive neurons. In addition to glutamatergic neurons in pain pathways, neurons in circuits involved in motivation and reward (VTA and nucleus accumbens), emotions (amygdala), and cognition (hippocampus and frontal cortex) also express high levels of mu opioid receptors. "Painful experiences are both personal and complex; they are not linearly correlated to noxious input but rather are constructed from neural information relating sensory, emotional, interoceptive, inferential, and cognitive information, which coalesce into a unified perception of pain" (Corder et al. 2018, 463).

The brain's motivation and reward circuits (figure 2.5) are intimately involved in addiction to opioids (Fields and Margolis 2015). "Reward" refers to the subjective feeling of pleasure. It involves changes in neuronal networks that increase the probability of a behavioral response that previously produced a beneficial or pleasurable outcome. The activation of mu opioid receptors and increased activity of dopaminergic neurons in the VTA are pivotal for the rewarding and addictive actions of morphine, heroin, and other opioids. To determine which neural circuits are involved in opioid addiction, researchers have placed small cannulas into different brain regions of rats. Small amounts of morphine or heroin flow through the cannula when the rat presses a lever in their cage. Rats will self-administer morphine when it is infused into the VTA or nucleus accumbens. Glenda Harris, Gary Aston-Jones, and their colleagues showed that infusion of drugs that block either AMPA or NMDA glutamate receptors into the VTA prevents morphine addiction in rats (Harris et al. 2004). Electrophysiological recordings from different types of neurons in the VTA have shown that opioids hyperpolarize GABAergic interneurons, which results in disinhibition of both glutamatergic and dopaminergic neurons.

It is very difficult for someone who becomes dependent on opioids to cease using them because withdrawal symptoms are often severe and prolonged. They include anxiety, intense cravings for the opioid, sweating, restlessness, nausea, fast heart rate, muscle pain, and insomnia. During withdrawal, there is a dramatic decrease in the activity of dopaminergic neurons in the VTA. The most commonly used method for helping addicts cease using opioids is to treat them with a weaker opioid called methadone. This treatment seems counterintuitive, however, and it turns out that methadone's action at mu opioid may not be the explanation for its effectiveness for opioid withdrawal. Instead, methadone has been shown to act as an NMDA receptor antagonist. Recent preclinical trials and case studies in people addicted to opioids suggest that the NMDA receptor antagonist ketamine can also lessen opioid-withdrawal symptoms.

PSYCHOSTIMULANTS

Psychostimulants are chemicals that increase one's ability to focus, improve mood, and reduce sleepiness. The caffeine in coffee and tea is the most commonly consumed psychostimulant. It enhances alertness and has no adverse effects on mental or physical health when consumed in moderate amounts daily. Caffeine enhances attention and cognition by inhibiting receptors for adenosine on glutamatergic neurons (Camandola, Plick, and Mattson 2019). Because activation of adenosine receptors suppresses cyclic AMP production, inhibition of these receptors by caffeine increases cyclic AMP levels in neurons. Chapter 4 described how by activating the transcription factor CREB, cyclic AMP enhances LTP at glutamatergic synapses and may thereby improve learning and memory. Caffeine can also increase the activity of dopaminergic neurons, which likely explains the pleasurable feeling it elicits. However, caffeine does not adversely affect most people's lives and so is not generally considered to be addictive.

Cocaine, methamphetamine, and MDMA exert their effects on glutamatergic neurons indirectly by inhibiting presynaptic dopamine transporters on dopaminergic neurons that innervate the glutamatergic neurons (Kalivas 2007). These drugs also inhibit, to varying amounts, serotonin and norepinephrine transporters. Methamphetamine is the most potent inhibitor of the dopamine transporter; cocaine comes next, and MDMA third. This explains why methamphetamine often causes a paranoid psychosis and why methamphetamine can cause degeneration of dopaminergic neurons and consequent symptoms like those of Parkinson's disease. Compared to cocaine and methamphetamine, MDMA has a relatively high affinity for serotonin transporters. This may explain why MDMA can induce hallucinations and why it can have an antidepressant effect. However, chronic abuse of MDMA can cause degeneration of serotonergic neurons and a constant state of anxiety.

Cocaine has been used in many regions of the world for more than 1,000 years. In the United States, cocaine was used as a local anesthetic in the 1800s, and only later as it became widely available did its high potential for abuse become clear. Many famous people have been addicted to cocaine. Some have been able to overcome their addiction, while others have died from cocaine abuse. It is well known that drugs of abuse are commonplace among people working in the entertainment industries. The actor and comedian John Belushi died at the age of 33 from an overdose of cocaine and heroin. Belushi's death caused his comedian friend Robin Williams to quit taking drugs. Unfortunately, about 20 years later Williams suffered from depression and relapsed to drinking alcohol and taking cocaine. The singer Whitney Houston is another example of someone who abused cocaine. At the age of 49, she was found dead in a bathtub. The autopsy reported that she died from cocaine use and an associated heart attack.

Some people have been able to overcome their cocaine addiction and lead productive lives. I recently watched the movie *The Chicago 7*, which was written and directed by Aaron Sorkin. Now 61 years old, Sorkin has had a remarkably productive career as a writer of Broadway plays such as *The Farnsworth Invention* and *A Few Good Men*, films such as *Moneyball* and *The Social Network*, and television series such as *The West Wing* and *The Newsroom*. Sorkin started using cocaine in 1987 and is reported to have said that it gave him relief from stress. He brought his addiction under control in 1995 but then in 2001 was arrested at an airport after security discovered cocaine in his carry-on bag. He has apparently not used the drug since then.

By causing dopamine accumulation at synapses, cocaine and methamphetamine engage evolutionarily important excitatory neuronal circuits. Activation of glutamatergic neuronal pathways that convey visual, auditory, olfactory, and other sensory information stimulate dopaminergic neurons in the VTA. These pathways evolved to enable mammals to respond to motivationally relevant novel objects and events in their environment (O'Connell and Hofmann 2011). The environmental stimulus may be either positive, such as a highly palatable food source or a potential mate, or negative, such as the presence of a predator or the occurrence of a thunderstorm. Experiments with rats and humans have shown that an initially unmotivating stimulus such as a tone or light can be made to trigger activation of dopaminergic neurons when it immediately precedes the motivational stimulus. From an evolutionary perspective, it is thought that this learning of the association between a rewarding or stressful stimulus and another environmental cue prepares the animal for an ensuing important event. Importantly, this neuronal circuitry evolved so as to adapt to the novel object or event such that the increased activation of dopaminergic neurons in response to the environmental stimulus decreases with repeated encounters with the stimulus.

Peter Kalivas has summarized the major differences between the activation of dopaminergic neurons that occurs during one's daily life and the activation that occurs in response to cocaine and methamphetamine:

> Since psychostimulants block the elimination of dopamine through DAT [a dopamine transporter], the level of dopamine achieved far exceeds what is possible from a biological stimulus. In contrast to biological stimuli that cease to release dopamine once an approach or avoidance response to that stimulus has been learned, psychostimulants continue to release large amounts of dopamine upon every administration. Thus, with psychostimulants, each administration releases dopamine into mesocorticolimbic regions, causing further associations to be made between the drug experience and the environment.

In this way, it is thought that the more a psychostimulant is administered, the more learned associations are made with the environment and the more effective the environment becomes at triggering craving and drug-seeking. It is this "overlearning" of drug-seeking behaviors by progressive associations formed between repeated drug-induced dopamine release and the environment that is thought to lead to increased vulnerability to relapse. (2007, 391)

The phenomenon of addiction to drugs involves both acute and long-term changes in neuronal circuits in multiple brain regions (Volkow and Morales 2015). The initial event upon drug administration is a robust release of dopamine from axon terminals of neurons whose cell bodies reside in the VTA. The axons of VTA dopaminergic neurons project to several different brain regions implicated in one or more features of addiction (figure 10.3): the nucleus accumbens, the dorsal and ventral striatum, the hippocampus, and the amygdala. Among these dopaminergic pathways, those from the VTA to the nucleus accumbens are critical for the initial rewarding effects of cocaine and methamphetamine and thus for the consequent binging and addiction. Long-term changes of glutamatergic synapses in projections from the prefrontal cortex to the nucleus accumbens, hippocampus, and amygdala are thought to mediate craving and the inability to overcome the urge to binge despite knowledge of the addiction's adverse effects on one's life. These synaptic adaptations may include insertion of AMPA receptors into dendritic spines.

After an addict ceases taking cocaine or methamphetamine, drug-seeking behaviors can be initiated once again by a stimulus previously associated with drug taking or by a new stressful situation. Drug seeking initiated by a previously paired stimulus results from dopamine release in the prefrontal cortex, and drug seeking initiated by a stressful stimulus results from dopamine release in the amygdala. The dopamine enhances the activity of glutamatergic neurons in the prefrontal cortex and amygdala, and those glutamatergic neurons in turn activate VTA dopaminergic neurons. Indeed, animal studies have shown that drug seeking is abrogated when glutamate receptors on VTA dopaminergic neurons are inhibited. It is thought that activation of glutamatergic neurons in the prefrontal cortex and amygdala mediate the addict's cognitive and emotional experiences, which manifest as craving.

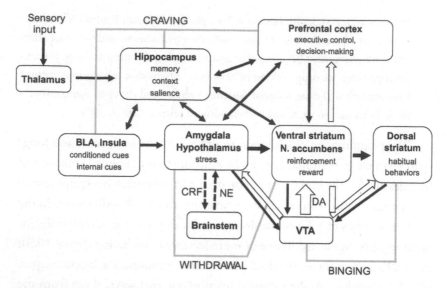

Figure 10.3

Neuronal networks involved in addiction. Binging occurs as the result of increased activity of dopaminergic neurons in the ventral tegmental area (VTA) and consequent release of dopamine onto medium spiny neurons in the nucleus accumbens as well as onto glutamatergic neurons in the prefrontal cortex. Craving involves increased activity of glutamatergic neurons involved in learning and memory as well as in goal-directed behaviors, including neurons in the hippo-campus, prefrontal cortex, amygdala, and insula. Craving is often triggered by sensory cues previously associated with drug ingestion, usually a place or person. Withdrawal symptoms result from activation of circuits involved in stress responses, including those in the brain-stem, amygdala, and hypothalamus. The solid-black arrows denote glutamatergic pathways. BLA = basolateral amygdala; CRF = corticotropin-releasing factor; DA = dopamine; NE = norepinephrine.

Relapse is very common among people who have quit using a psycho-stimulant or other addictive drug. Animal studies have shown that during drug seeking there occurs a large release of glutamate from axon terminals of neurons that project from the prefrontal cortex to the nucleus accumbens. This increased glutamate release likely mediates relapse because drugs that inhibit the glutamate release also inhibit drug seeking.

Animal studies have shown that chronic cocaine administration (over several weeks or months) results in the structural remodeling of neuronal circuits in multiple brain regions. Such remodeling includes an increased density of dendritic spines at glutamatergic synapses on medium spiny neurons in the nucleus accumbens and increased dendritic complexity and synapse numbers on glutamatergic pyramidal neurons in the prefrontal cortex (Robinson and Kolb 1999). It is thought that these enduring structural changes are responsible, at least in part, for the addictive actions of amphetamines and other drugs.

Tobacco use remains a leading cause of disability and death throughout the world. People who regularly smoke tobacco cigarettes, cigars, or pipes are at increased risk for lung cancer, cardiovascular disease, and hypertension. Use of chewing tobacco increases one's risk for throat and esophageal cancer. Tobacco products are addictive because they contain nicotine. As with other addictive drugs, the acute effects of nicotine are pleasurable and include increased alertness, mild euphoria, and decreased fatigue. Repeated use of nicotine results in structural and functional adaptations of the brain's reward circuits. Cessation of nicotine results in negative symptoms, including anxiety, difficulty concentrating, irritability, and craving. Relapse often occurs and can be triggered by environmental cues previously associated with nicotine intake, such as a specific location, a friend who smokes, and so on.

Nicotine activates acetylcholine receptors that are widely distributed in neurons throughout the brain. The nicotinic acetylcholine receptors are ion channels that flux Na^+ and Ca^{2+}. Binding of nicotine to receptors on the presynaptic terminals of glutamatergic neurons results in Na^+ influx, membrane depolarization, and Ca^{2+} influx through the nicotine receptor channel as well as through voltage-dependent Ca^{2+} channels. The activity of VTA dopaminergic neurons is increased by inputs from glutamatergic neurons in the amygdala and prefrontal cortex. Electrophysiological recordings from VTA dopaminergic neurons have shown that nicotine enhances both glutamate release and activation of AMPA and NMDA receptors (D'Souza and Markou 2013; Pistillo et al. 2015). When rats learn to chronically self-administer nicotine, there occurs an increase in the amount of AMPA receptors in their VTA neurons. When AMPA or NMDA receptor antagonists are infused into the VTA, dopamine release is reduced. Moreover, administration of NMDA receptor antagonists decreases nicotine self-administration in rats. Genetic-engineering methods have been used to produce mice that lack NMDA receptors in VTA dopaminergic neurons. These mice exhibit reduced drug-seeking behaviors in response to cues associated with nicotine administration. Altogether, the evidence suggests that glutamatergic inputs to the VTA play fundamental roles in the processes of nicotine addiction and withdrawal symptoms.

GABAergic neurons also contain presynaptic nicotinic receptors, and nicotine can increase GABA release. However, evidence suggests that the sensitivity of GABAergic neurons to nicotine diminishes with repeated exposures. In combination with the increase in glutamatergic activity, the desensitization of GABA receptors likely enhances the activity of VTA dopaminergic neurons (D'Souza and Markou 2013).

The results of studies of animal models of nicotine addiction suggest that drugs that inhibit glutamate receptors or activate GABA receptors can be beneficial in helping people cease use of tobacco. However, there have been only a few clinical trials of such drugs in people addicted to nicotine. In one study, Teresa Franklin and her colleagues at the University of Pennsylvania performed a double-blind placebo-controlled trial of the GABA-B receptor agonist baclofen in 60 smokers. During a nine-month period, the smokers receiving baclofen significantly reduced the number of cigarettes they smoked each day and reported less craving compared to those in the placebo group (Franklin et al. 2009).

ALCOHOL

Humans have consumed alcohol for at least 10,000 years. Soon after consuming one or two alcoholic beverages, a person will experience happiness, decreased anxiety, increased sociability, and euphoria. They will also experience impaired cognition, drowsiness, and impaired sensory and motor function. Consumption of higher amounts of alcohol may cause dizziness, nausea, and loss of consciousness.

Regular consumption of high amounts of alcohol often results in dependence and is a risk factor for spousal abuse, depression, and suicide. When someone who has become dependent stops ingesting alcohol, they will experience severe anxiety, shivering, and confusion. They may also have seizures.

Upward of 15 million people in the United States have an alcohol-use disorder, and each year approximately 100,000 die as a result of excessive intake of alcohol. About half of those deaths are due to diseases caused by chronic alcohol use, including liver disease, heart disease, and cancers. Others die from alcohol-intoxication-related accidents and violence. By exacerbating respiratory suppression, alcohol also contributes to many overdose deaths involving opioids and benzodiazepines.

It is thought that alcohol's acute sedative and anxiolytic effects result mainly from enhancement of GABAergic signaling. As with other addictive drugs, alcohol's rewarding effects, including pleasure and increased sociability, result from increased activity of dopaminergic neurons in the VTA (Alasmari et al. 2018). The results of fMRI studies in which brain scans are performed on people within an hour of consuming alcohol are generally consistent with alcohol's suppression of cerebral cortical networks involved in cognition, sensory and motor functions, as well as its activation of the mesolimbic dopaminergic system (Bjork and Gilman 2014; Dupuy and Chanraud 2016).

Animal studies have shown that alcohol dependence is associated with increased dopamine release in the prefrontal cortex and amygdala. In turn, there is increased activity of glutamatergic neurons in layer V of the prefrontal cortex that project to the nucleus accumbens, hippocampus, and amygdala. Activation of those glutamatergic neurons enhances the rewarding effects of the alcohol. As Peter Kalivas and Nora Volkow describe this effect,

> Pathophysiological plasticity in excitatory transmission reduces the capacity of the prefrontal cortex to initiate behaviors in response to biological rewards and to provide executive control over drug seeking. Simultaneously, the prefrontal cortex is hyperresponsive to stimuli predicting drug availability, resulting in supraphysiological glutamatergic drive in the nucleus accumbens, where excitatory synapses have a reduced capacity to regulate neurotransmission. Cellular adaptations in prefrontal glutamatergic innervation of the accumbens promote the compulsive character of drug seeking in addicts by decreasing the value of natural rewards, diminishing cognitive control (choice), and enhancing glutamatergic drive in response to drug-associated stimuli. (2005, 1403)

People who become addicted to a drug typically go through cycles of abstinence and relapse. When alcohol is withdrawn from rats that had become addicted, an increase in glutamate release in the nucleus accumbens occurs. Evidence suggests that this glutamate release is due in part to increased activity of glutamatergic neurons in the prefrontal cortex that project to the nucleus accumbens. Neuronal circuits in the prefrontal cortex are thought to store and process information not currently in one's immediate environment in ways that control goal-directed behaviors. In the case of drug addiction, the behavior is drug seeking. In addition, glutamatergic neurons in the amygdala that project to both the nucleus accumbens and the prefrontal cortex are implicated in drug seeking and relapse. Circuits in the amygdala are important for applying an emotional context to alcohol—the pleasurable experience of alcohol ingestion and the distressing experiences of withdrawal. Hippocampal circuits are responsible for remembering when, where, and with whom alcohol was consumed previously. Findings from animal studies suggest that glutamatergic projections from the hippocampus to the prefrontal cortex and nucleus accumbens are involved in relapse to the use of alcohol and other addictive drugs.

The process of addiction can be divided into three recurring stages—binging, withdrawal, and craving (figure 10.3). Over time, this three-stage cycle worsens in terms of its adverse effects on mental and physical health. Functional and structural changes in the brain's reward circuits mediate binging behavior. Changes in circuits that mediate stress responses also mediate the addict's emotional experiences during withdrawal, while changes in circuits involved in memory and executive control are responsible for craving.

Brain-imaging studies have shown that chronic heavy drinkers have enlarged cerebral ventricles and reductions in overall brain size and gray matter volumes in the cerebral cortex, hippocampus, and cerebellum (Le Berre et al. 2014; X. Yang et al. 2016). Some studies have shown that the extent of gray matter loss is directly correlated with the number of years of heavy drinking. Longitudinal studies have shown that people who have overcome their alcohol dependence exhibit a recovery of gray matter loss compared to when they were addicted to alcohol, like what occurs in people who have recovered from major depression.

There is a major need for drugs that can help alcoholics fully recover without recurrence of the addiction. The results of a recent study reported by Elias Dakwar, Edward Nunez, and their colleagues at Columbia University suggest that the NMDA receptor antagonist ketamine can help alcoholics recover. They found that a single dose of ketamine significantly reduced the number of heavy-drinking days and increased abstinence (Dakwar et al. 2020). As mentioned earlier, psychedelics have also been reported to be beneficial in helping people recover from alcoholism. It remains to be seen whether ketamine and psychedelics become common treatments for people addicted to alcohol or other drugs.

BENZODIAZEPINES

Benzodiazepines are used to treat anxiety, insomnia, and epileptic seizures. These drugs activate type A GABA receptors and thereby suppress the activity of the glutamatergic neurons they innervate. The most commonly prescribed benzodiazepines are diazepam, alprazolam, clonazepam, and lorazepam. They are safe and effective for short-term treatment, but long-term use results in the development of tolerance and thus to addiction. Indeed, following opioids, benzodiazepines are the second type of prescription drug to which people most commonly become addicted.

As with other addictive drugs, benzodiazepines increase the activation of VTA dopaminergic neurons (Tan, Rudolph, and Luscher 2011). At first approximation, this effect seems counterintuitive. How does activation of inhibitory GABA receptors increase the activity of dopaminergic neurons? Evidence from animal studies suggests the following answer. GABAergic neurons themselves have GABA receptors. Benzodiazepines potently inhibit the GABAergic inputs to the VTA dopaminergic neurons while having little direct inhibitory effect on the dopaminergic neurons. This cellular mechanism is called "disinhibition." Subsequent to this initial effect of benzodiazepines on GABAergic neurons, the neuronal circuits involved in addiction, withdrawal, and relapse are believed to be similar if not identical to those engaged by other addictive drugs.

WHAT ABOUT WEED?

The plant *Cannabis sativa*—commonly referred to as "marijuana"—contains the psychoactive chemical THC. Strains of *C. sativa* with low amounts of THC are called "hemp" and are cultivated for use in the production of hemp fiber, seeds, and oils. Archaeological and gene-sequencing data indicate that cannabis cultivation began about 12,000 years ago in eastern Asia. Its use for mind-altering purposes is documented in Greek historical documents dating to about 2,400 years ago. In the United States, it is legal to cultivate *C. sativa* that contains no more than 0.3 percent THC. Although federal law prohibits the possession of marijuana with higher levels of THC, an increasing number of states have legalized its possession and regulate its cultivation and sales. Strains with relatively high amounts of THC are used for both recreational and medical purposes.

Next to alcohol and tobacco, marijuana is the most commonly used psychoactive drug that has a potential for addiction. It has been estimated that more than 200 million people throughout the world regularly use marijuana. When marijuana is smoked, THC rapidly enters the blood and brain. Within minutes, the THC affects neuronal networks throughout the brain, resulting in a wide range of effects on cognition, mood, and behavior, such as euphoria, heightened senses, reduced anxiety, reduced pain perception, and disinhibition in social interactions. These effects can last for several hours. However, during this period of intoxication, cognition and motor skills are impaired, and one's sense of time can be distorted. The latter effects of marijuana can lead to accidents and can impair performance in school and work. Most people who partake of marijuana, as of alcohol, do so in limited amounts and do not become addicted. Nevertheless, studies have consistently shown that marijuana abuse is a major problem for many adolescents and adults.

The mind-altering and addictive effects of marijuana are mediated by THC acting on the cannabinol 1 (CB1) receptors on neurons (Cristino,

Bisogno, and Marzo 2020). The CB1 receptors are present throughout the brain and are particularly abundant in the hippocampus, cerebral cortex, basal ganglia, and cerebellum. These receptors interact with a GTP-binding protein on the inner membrane surface and are concentrated in presynaptic terminals of GABAergic and glutamatergic neurons. Binding of THC to the CB1 receptor results in the inhibition of voltage-dependent Ca^{2+} channels and a consequent inhibition of neurotransmitter release from the presynaptic terminal.

The discovery that THC is the main psychoactive component of marijuana was made in the mid-1960s, and the CB1 receptor and its signaling mechanism were discovered in the 1990s. The question then became: What is the endogenous signaling molecule that activates the CB1 receptor? Two neurochemicals were found to selectively bind to the CB1 receptor, anandamide and 2-arachidonylglycerol. These endogenous ligands for CB1 are called "endocannabinoids" (Wilson and Nicoll 2002). Unlike classical neurotransmitters, which are or are derived from amino acids, endocannabinoids are lipids derived from arachidonic acid, a component of cell membranes. Studies have shown that glutamate is the major stimulus for the production of endocannabinoids in the brain. Activation of AMPA and NMDA receptors results in Ca^{2+} influx, and the Ca^{2+} then activates enzymes that generate endocannabinoids. Also in contrast to classical neurotransmitters, the endocannabinoids are not packaged in vesicles but instead diffuse retrogradely from their site of production in the postsynaptic membrane to presynaptic terminals, where they activate CB1 receptors and so suppress neurotransmitter release.

The first investigations of the cellular localization of CB1 demonstrated that it was highly concentrated in the presynaptic terminals of GABAergic neurons (Colizzi et al. 2016). Electrophysiological recordings from glutamatergic neurons in hippocampal slices showed that activation of CB1 inhibits GABA release and thereby increases the excitability of the glutamatergic neurons. This phenomenon has been termed "depolarization-induced suppression of inhibition." Further findings suggested that CB1 is present in high amounts in a subset of inhibitory interneurons that have particularly fast kinetics of GABA release. This fast release of GABA is a consequence of rapid influx of Ca^{2+} through a particular type of voltage-dependent Ca^{2+} channel called the "N channel." Depolarization-induced suppression of inhibition enhances LTP at glutamatergic synapses, and this effect is abolished by drugs that block CB1 receptors. Based on the later findings, one might speculate that marijuana may improve cognition because LTP is associated with learning and memory. However, this is not true because learning and memory are not impaired in mice lacking CB1, and human studies have shown that marijuana impairs performance on cognitive tests for at least several hours after ingestion.

Although early findings suggested that CB1 receptors are located primarily in the presynaptic terminals of GABAergic neurons, more recent evidence suggests that they are also present in the presynaptic terminals of glutamatergic neurons (Colizzi et al. 2016). Electrophysiological studies have shown that THC inhibits the release of glutamate at synapses on the dendrites of CA1 pyramidal neurons. This effect of THC is thought responsible for the cognitive impairment that occurs during marijuana intoxication. However, because CB1 receptors are widely distributed in both GABAergic and glutamatergic neurons throughout the brain, marijuana affects many different processes necessary for performance on cognitive tests, including attention, sensory processing, and motivation.

Each GABAergic interneuron forms synapses with hundreds of glutamatergic neurons. Studies have shown that this extensive connectivity plays an important role in synchronizing the activity of glutamatergic neuronal networks. Electrophysiological recordings have provided evidence that activation of CB1 receptors suppresses the synchronized firing of glutamatergic neurons (Cortes-Briones et al. 2015). GABAergic neurons are thought to orchestrate the synchronized oscillations of cerebral cortical neuronal network activity in the high-frequency "gamma" range. These gamma oscillations are evident in EEG recordings. It therefore can be predicted that activation of CB1 receptors on GABAergic presynaptic terminals would have widespread effects on synchronized neuronal network activities. Results from human studies have indeed shown that soon after administration of THC, gamma oscillations are disrupted. It remains to be determined whether this effect of THC plays a role in the effects of marijuana on mental states.

Cannabis can be addictive. Studies have shown that activation of CB1 by THC increases dopamine release from VTA neurons. The magnitude of the enhancement of dopamine release by THC is considerably less than that of opioids, cocaine, and methamphetamine, which may explain why the latter drugs of abuse are more addictive than cannabis. In contrast to the direct effect of cocaine and methamphetamine on dopaminergic neurons, the enhancement of dopamine release by THC is indirect. Animal studies have shown that THC inhibits glutamate release onto GABAergic neurons in the nucleus accumbens, which results in a reduction in inhibition of the VTA dopaminergic neurons by the nucleus accumbens GABAergic inputs.

Evidence suggests that regular cannabis use can have particularly untoward effects on the brains of adolescents (Lorenzetti, Hoch, and Hall 2020). Adolescents who regularly use marijuana perform more poorly on IQ and verbal-learning tests compared to those who do not use cannabis. They are also at an increased risk of developing schizophrenia. Studies suggest that cognitive deficits caused by marijuana use can be reversed with sustained abstinence.

Justine Renard, Steven Laviolette, and their coworkers at the University of Western Ontario exposed adolescent rats to either THC or placebo twice daily for eleven days and then recorded activities of glutamatergic neurons in the prefrontal cortex and of dopaminergic neurons in the VTA. They found that activities of both the glutamatergic and dopaminergic neurons was increased in the rats treated with THC (Renard et al. 2017). The THC-treated rats exhibited impaired short-term memory, increased anxiety, and impaired motivation for social interactions with other rats. Infusion of the GABA receptor activator muscimol into the prefrontal cortex reversed the adverse effects of THC on VTA dopaminergic activity, normalized short-term memory and social interactions, and reduced anxiety levels. These findings suggest an important role for alterations in the activities of prefrontal cortex GABAergic and glutamatergic neurons in the adverse effects of marijuana use on the adolescent brain.

Cannabis is increasingly used to treat several medical problems. It has been demonstrated to be effective in reducing nausea resulting from chemotherapy in cancer patients. Although limited in number, clinical trials suggest that cannabis and THC may also be beneficial for patients with chronic pain and multiple sclerosis (Chaves, Bittencourt, and Pelegrini 2020; Fragoso, Carra, and Macias 2020). Several studies have reported reductions in pain and improvements in function in people with chronic neuropathic pain by means of THC. One brain-imaging study provided evidence that the reduction in pain perception in response to THC administration was correlated with a reduction in functional connectivity between the sensory cortex and the anterior cingulate cortex (Weizmann et al. 2018). The latter findings are consistent with evidence that glutamatergic projections from sensory cortex to anterior cingulate cortex mediate one's perception of pain. As with psychedelics and ketamine, there has recently been a marked increase in the number of clinical trials studying the use of cannabis for multiple brain disorders. Whether these drugs will be incorporated into mainstream medicine remains an open question.

11 SCULPTING A BETTER BRAIN

An evolutionary perspective informs us how the brains of all animals, including humans, came to be as they are. Evolutionary considerations can also provide insight into how the environment influences the structure and function of the brains of individuals living today. This chapter considers how three evolutionarily important environmental challenges can improve brain function and resilience by enhancing the plasticity of glutamatergic neuronal networks: food deprivation/fasting, physical exercise, and intellectual challenges.

Evidence suggests that food scarcity was the major driver of brain evolution (Mattson 2022). When viewed in the light of evolution, brains and bodies clearly changed so as to function well and perhaps optimally in a food-deprived state. This is evident in the lives of predators. For example, wolves often go for many days or even weeks without killing a prey animal. During such periods of food deprivation, a wolf's brain and body must be capable of functioning very well for it to be successful in finding and killing a prey animal. Individuals with traits that enhanced their success in food acquisition survived and passed their genes on to the next generation. One such trait in wolves is cooperation with other members of a pack to capture and kill a large prey animal that any individual wolf could not kill by itself. Similarly, humans' "social brain" evolved as an adaptation enabling cooperation in food acquisition, processing, and distribution.

As with most other animals, our human ancestors lived in environments where food sources were often scarce, varying with weather and seasons. In

The Intermittent Fasting Revolution (2022), I proposed that even the most advanced capabilities of the human brain, including creativity, imagination, and invention, originally evolved as adaptations that enabled success in food acquisition and maximizing the nutritional value of the foods. All of the early tools invented by humans were for the purposes of acquiring and processing food: flaked stones for cutting, spears, bow and arrow, animal traps for killing wild animals; stones, mortar and pestle, and fire for processing foods. Small groups of hunter-gatherers proliferated and began to move out of Africa. Then about 11,000 years ago, humans learned to domesticate animals for meat, including sheep, cattle, and goats, and shortly thereafter they domesticated certain grain plants, including wheat and barley in the fertile crescent of what is now the Middle East and rice in what is now Southeast Asia.

The ability to produce and store large amounts of food enabled the establishment of small settlements and then larger cities and countries. Horse-powered farming equipment and wagons facilitated the dispersal of agricultural products. Because an increasingly smaller percentage of the people were able to produce enough food for the entire population, others could work in other specialized occupations, such as teaching, carpentry, blacksmithing, clothes making, engineering, medicine, and so on. Governments were established, and laws were enacted and enforced. Brain neuronal networks that originally evolved to solve the problem of food scarcity were repurposed for such occupations.

There is considerable evidence that intellectual challenges enhance the neuroplasticity of glutamatergic circuits. For example, people who engage regularly in intellectual challenges are less likely to develop cognitive impairment and Alzheimer's disease as they age (Scarmeas and Stern 2004). This finding supports the concept of "use it or lose it." Like muscle cells during exercise, when neurons are stimulated by glutamate during intellectual challenges, they experience ionic, metabolic, and oxidative stress. The neurons respond to this stress adaptively in ways that not only improve their functionality but also enhance their ability to cope with stress and resist disease. Cellular stress-associated signals such as Ca^{2+} and free radicals mediate adaptive responses by neurons to intellectual activity.

Although it is intuitively obvious that the brain's intellectual capabilities can be improved by engaging in intellectual challenges, it is perhaps less obvious that physical exercise and intermittent fasting can also improve cognition. Data from animal and human studies do show that exercise and moderation in calorie intake are also associated with better cognition as well as improved mood. This chapter considers how the challenges of intellectual endeavors, physical exercise, and eating patterns that include periods of fasting affect the plasticity of glutamatergic circuits in ways that enhance cognition, improve mood, and possibly protect the brain against a wide array of disorders, including epilepsy, stroke, Alzheimer's disease, and Parkinson's disease.

This chapter also describes how increasingly sedentary and overindulgent lifestyles adversely affect the brain throughout the life course and across generations. For example, brain-imaging studies have consistently demonstrated reduced sizes of the hippocampus in people with obesity compared to people with lower body weights. This is true in children as well as in adults. Animal studies have elucidated how obesity and diabetes adversely affect the structure and function of glutamatergic neuronal networks. These kinds of findings are of concern in light of the dramatic increase in the prevalence of obesity and diabetes in most industrialized countries of the world today.

USE IT OR LOSE IT

Epidemiological studies have shown that people who regularly engage in intellectually challenging endeavors throughout their lives are less likely to develop dementia as they age compared to people who lead less cognitively stimulating lives (Bielak 2010). Regular participation in activities such as reading, attending lectures, playing chess, and experiences that require considerable decision-making are associated with slower age-related declines in their ability to compare figures or symbols quickly and accurately or to find figures or symbols in lists. Such intellectual challenges involve activity in and structural plasticity of the glutamatergic neurons that encode and process incoming and stored information.

I am now in my midsixties and have begun to notice a slowing of my own information-processing speed. During usual daily activities, I am unaware of this effect of aging on the glutamatergic neuronal networks in my brain. However, it becomes very evident when I am challenged with answering questions quickly, particularly when the questions require integration of multiple bits of information. For example, lately I have been watching the British quiz program *University Challenge*, a competition where four-person teams of students from colleges compete against each other in a double-elimination battle of wits. Of course, they all are very smart. The questions are in areas ranging from physics, biology, and chemistry to history, the arts, and literature. A typical question might be: "Noting that in this respect he was almost alone among the philosophers, of whom did Hannah Arendt say he was much bothered by the common opinion that philosophy is only for the few precisely because of its moral implications. The philosopher was born in 1784." Rapid cognitive-processing speed is essential to answer the questions quickly. Even in instances where I do know the answer to a question, the students are usually much quicker in coming up with it.

The question for us here then becomes: What are the mechanisms by which engaging in such intellectual challenges can sustain cognition during aging? Not surprisingly, the answers are emerging from studies of glutamatergic circuits in the hippocampus that mediate learning and memory processes.

Functional brain-imaging studies have shown that neuronal activity increases in the hippocampus when people are engaged in cognitive tasks (Nees and Pohlack 2014). Compared to mice or rats living in cages without objects to explore, those living in cages with objects exhibit better learning and memory in a variety of tests. In particular, animals in such enriched environments are able to retain memories for longer periods of time compared to those living in the control environment (Leger et al. 2012). Mice in the enriched environments exhibit increased activation of glutamatergic neurons in the hippocampus and frontal cortex. Environmental enrichment has been shown to increase the numbers of synapses on the dendrites of pyramidal neurons and dentate gyrus granule neurons in the hippocampus (Ohline and Abraham 2019).

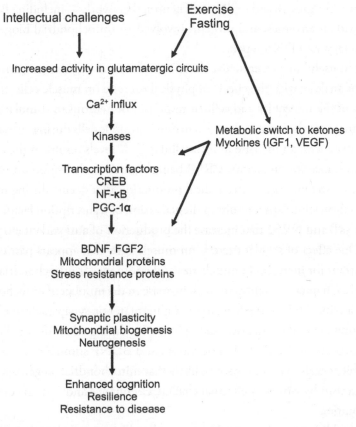

Intellectual challenges   Exercise
Fasting

↓   ↙

Increased activity in glutamatergic circuits

↓

Ca²⁺ influx

↓

Kinases

Metabolic switch to ketones
Myokines (IGF1, VEGF)

↓

Transcription factors
CREB
NF-κB
PGC-1α

↓

BDNF, FGF2
Mitochondrial proteins
Stress resistance proteins

↓

Synaptic plasticity
Mitochondrial biogenesis
Neurogenesis

↓

Enhanced cognition
Resilience
Resistance to disease

Figure 11.1

Mechanisms by which intellectual challenges, exercise, and intermittent fasting affect gluta-matergic neuronal circuits in ways that enhance cognition and the brain's resistance to stress and disease.

Chapter 4 described how learning and memory involves the influx of Ca^{2+} into dendrites that occurs in response to the activation of glutamate receptors. This Ca^{2+} influx initiates a coordinated sequence of molecular events that result in long-term structural and functional adaptations of the neuronal connections (figure 11.1). These events include the phosphorylation of glutamate receptor proteins, increased insertion of AMPA receptors into the postsynaptic membrane, the generation of nitric oxide, the activation of the transcription factors, and the production of proteins that promote the growth and resilience of neuronal networks—including BDNF, antioxidant enzymes, and proteins involved in mitochondrial biogenesis, autophagy, and DNA repair.

It is useful to conceptualize the effects of mental exercise on neurons by way of analogy with the effects of physical exercise on muscle cells. In fact, many of the molecular and cellular responses of neurons to stimulation by glutamate are the same as those occurring in muscle cells during exercise. In both instances, an increase in intracellular Ca^{2+} levels results in the activation of transcription factors, CREB being a particularly important one. In addition, the increase in free radical production that occurs during muscle and neuron stimulation results in the activation of transcription factors such as NF-κB and NRF2 that increase the production of antioxidant enzymes.

One effect of regular exercise on muscle cells that appears particularly important for increases in muscle size and endurance is mitochondrial biogenesis. Chapter 5 described how an increase in the number of mitochondria plays a critical role in the formation of glutamatergic synapses during brain development and the maintenance of synapses during adult life. Intellectual challenges stimulate BDNF production, and BDNF stimulates mitochondrial biogenesis. It therefore seems likely that mitochondrial biogenesis is one mechanism by which intellectual challenges enhance and sustain cognitive capabilities.

In addition to enhancing the structural and functional plasticity of extant neuronal circuits, intellectual challenges may also stimulate hippocampal neurogenesis. The proliferation of neuronal stem cells in the hippocampus is increased in rats and mice that are maintained in enriched environments (Kempermann 2019). The newly generated neurons become granule neurons that integrate into the hippocampal circuitry. The results of studies in which neurogenesis is prevented suggest that neurogenesis plays a particularly important role in the learning of spatial information, as in a mouse remembering the route through a maze to a food source. Enriched environments, running, and intermittent fasting—all three have been shown to enhance hippocampal neurogenesis. These bioenergetic challenges increase activity of glutamatergic neurons in the hippocampus. Working in Singapore and Norway, Ingrid Amellem, Ayumu Tashiro, and their colleagues recently showed that activation of NMDA receptor in dentate granule neurons is required for normal proliferation and differentiation of stem cells (Amellem et al. 2021). FGF2 and BDNF are produced in response to activation of glutamate receptors. FGF2 stimulates neuronal stem cell proliferation, and BDNF promotes the growth and survival of newly generated neurons. The increased activity of dentate granule neurons that occurs during intellectual challenges increases the production of these neurotrophic factors.

Interactions with natural environments may have even greater beneficial effects on cognition and mood than interactions with artificial environments. This has been shown in studies in which responses to stress are measured when people are put in natural or artificial environments (Mostajeran et al. 2021). After watching a stress-provoking movie, subjects who had previously viewed natural scenes recovered more quickly than did those who had viewed urban scenes. There is considerable evidence that living in environments in which our species did not evolve, such as large cities, is causally linked to increases in anxiety disorders and depression. The question then becomes: How do natural and artificial environments differentially affect neuronal networks involved in stress responses and mood?

Kelly Lambert and colleagues asked whether environments enriched with natural and environments enriched with artificial objects would have the same or different effects on neuronal network activities and behavior in rats (Lambert et al. 2016). Rats were divided into three environments: cages with no objects to explore; cages with artificial objects such as balls, plastic shelters, slinkies, and so on; and cages with natural objects such as sticks, rocks, hollow logs, and so on. The rats in the environment with natural objects spent more time interacting with the objects compared to those in the environment with artificial objects. Interestingly, rats in the environment with natural objects also spent more time interacting with each other. Evaluation of the rats' glutamatergic neuronal network activity revealed reduced activity in neurons in the amygdala and more activity in neurons in the nucleus accumbens in rats in the environment enriched with natural objects. This makes sense because activation of glutamatergic neurons in the amygdala increases anxiety, whereas activation of glutamatergic neurons in the nucleus accumbens occurs in response to rewarding experiences.

MOTION IS BRAIN LOTION

If you exercise regularly, you are well aware of its beneficial effects on your mood and cognition. This is most striking when for some reason you are unable to exercise—your mood will plummet, and you may notice a decrement in your ability to concentrate on a task for extended periods of time. Like many other people, I have experienced both the benefits of exercise for mental clarity and mood as well as the adverse effects on mental health of not being able to exercise because of injuries. These perceived effects of exercise on cognition and mood have been confirmed in controlled studies in humans, and animal studies have elucidated the underlying cellular and molecular mechanisms.

Many of the human studies have been randomized controlled trials of aerobic exercise, with the control group doing only light stretching and toning. Stanley Colcombe, Arthur Kramer, and colleagues at the University of Illinois reported that the volumes of several brain regions—including the supplementary motor area, the anterior cingulate cortex, and the hippocampus—were increased during a six-month period of aerobic exercise (Colcombe et al. 2006). Subsequent studies documented increases in the size of the hippocampus in response to aerobic exercise. In a recent study of 132 cognitively average people ages 20 to 67, Yakaav Stern and colleagues at Columbia University found that a six-month period of aerobic exercise significantly improved working memory compared to just stretching and toning (Stern et al. 2019). Interestingly, the effect of aerobic exercise on cognition was greater in participants who were in their fifties and sixties compared to the younger participants. In the same study, the participants' brains were imaged with MRI before and at the end of the exercise period, and the thickness of the cerebral cortex was measured from the images. There was an increase in cortical thickness in the aerobic exercise group but not in the control group.

As alluded to in chapters 4 and 5, functional brain imaging provides an indirect measure of the activity of glutamatergic neurons because those neurons constitute more than 90 percent of the neurons in the brain, and their activity is coupled to the blood flow measured by fMRI. Studies have documented that the effects of aerobic exercise on neuronal network activity differ among brain regions and cognitive tests. Cynthia Krafft and colleagues at the University of Georgia performed cognitive testing and fMRI analysis of overweight children 8 to 11 years old before and after an eight-month exercise program (Krafft et al. 2014). The exercise group played tag and jumped rope for 40 minutes per day, five days each week. The control group played board games and made artwork for 40 minutes per day, five days each week. The children's cognition was evaluated using a standardized series of tests that assessed their attention, planning ability, and simultaneous and successive processing. These tests included a "flanker task" and an "antisaccade task." For the flanker task, the children were presented with sequences of letters or symbols that are either congruent, such as >>>>>, or incongruent, such as >><>>. The children were instructed to press the left button if the middle symbol in the sequence was the same as the flanking (surrounding) sequences and the right button if the middle symbol was different than the flanking symbols. Response time and accuracy were recorded. For the antisaccade task, the children were asked to focus on a motionless object. A stimulus was then presented to one side of the object, and the children were asked to look to the other direction. Failure to look away from the stimulus was considered an error.

There were no significant differences between exercise and control groups in the antisaccade test, which may be due to this test's relative simplicity. Nevertheless, fMRI data showed that children in the exercise group had decreased neuronal network activity in the precentral gyrus and posterior parietal cortex, brain regions known to be involved in antisaccade movements. In contrast, activities in neuronal networks involved in flanker task performance, including the superior frontal gyrus and anterior cingulate cortex, were greater in children in the exercise group. These differences were correlated with aerobic fitness as measured by "VO_2 max" (the volume of oxygen your body can absorb during exercise) on a child-by-child basis.

Studies using cognitive tests that are more difficult than the flanker or antisaccade tasks have demonstrated that exercise can improve learning and memory. Such studies, which typically include aerobic exercise for 30 minutes or more at least three days per week, have shown that exercise is particularly effective in improving working memory and executive function.

Chapter 4 described working memory and executive functions. A commonly used test of working memory is the N-back test. In this test, the participant is shown cards with numbers or letters and then asked to name the letter or number on a previous card. For example, the participant is shown cards with the numbers 7, 8, 5, 3, and 4 and then asked to name the number shown three cards before the last card. The correct answer is 8. The difficulty is increased by increasing the number of cards shown. This test determines the participant's ability to remember sequences of events. Tests for executive function include the Stroop test and the trail-making test. For the Stroop test, the participant is shown a card with the name of a color printed in letters of a color different from the named color. For example, the word *red* is spelled out in green letters. The participant is asked to say the color spelled by the letters. The cards are presented in rapid succession. Then the participant is shown the cards again and asked to name the ink color instead of the word spelled by the letters. Errors are recorded. For the trail-making test, the participant is asked to connect a sequence of 25 numbers and/or letters on a sheet of paper in order. They are then given a sheet of paper with the numbers 1 to 13 and the letters A to L and asked to connect them in order while alternating letters and numbers and to do it as quickly as possible. Functional brain-imaging studies suggest that glutamatergic networks in the prefrontal cortex and hippocampus play major roles in such cognitive abilities (L. Li et al. 2014).

Many published studies have documented beneficial effects of regular aerobic exercise on tests of learning and memory like those described in the previous paragraph. Knowledge of the cellular and molecular mechanisms by which exercise improves cognition has come from animal studies (Mattson 2012; Voss et al. 2019). The most common exercise studies in animals have compared rats or mice with running wheels in their cages to those without running wheels. Rats and mice will often run more than five miles every day, and they do so in short bouts of several seconds to a few minutes. Running-wheel exercise improves the performance of rats and mice in tests of spatial learning and memory that depend on the hippocampus. Such tests involve navigation through a maze. The effects are particularly robust when comparing older animals that had running wheels in their cages throughout their adult life with control animals that had no running wheels in their cages.

Electrophysiological recordings of glutamatergic synaptic activity in slices of hippocampus from runner and sedentary animals have shown that exercise increases the strength of glutamatergic synaptic transmission (Farmer et al. 2004), which may result from a combination of increased numbers of synapses and increased size of synapses. Exercise also enhances neurogenesis in the hippocampus by stimulating the proliferation and differentiation of neuronal stem cells. Working at the US National Institute on Aging, Carmen Vivar and Henriette van Praag showed that the new dentate granule neurons produced in response to exercise receive synaptic inputs from neurons in several other brain regions, including the entorhinal cortex, the medial septum, and mammillary bodies (Vivar et al. 2012). These brain regions are important for encoding spatial and temporal information and for remembering the context of experiences.

Evidence suggests that several neurotrophic proteins—including BDNF, FGF2, and IGF1—play important roles in the beneficial effects of exercise on synaptic plasticity, neurogenesis, and learning and memory (Cotman and Berchtold 2002; Mattson 2012). Levels of each of these neurotrophic factors increase in the hippocampus in response to exercise. The mechanism by which exercise stimulates the production of BDNF and FGF2 involves increased activity in hippocampal neurons during the exercise and may also involve changes in energy metabolism, such as increased lactic acid production. Chapter 4 described how BDNF increases the size and numbers of synapses that occur in response to intellectual challenges. This may also be true for the cognitive enhancement afforded by exercise. In addition, BDNF plays an important role in neurogenesis, where it promotes the survival and growth of newly generated neurons as well the new neurons' integration into dentate gyrus circuits.

FGF2 stimulates the proliferation of neuronal stem cells and enhances the growth of glutamatergic neurons' axons and dendrites. Studies of mice genetically engineered to lack the receptor for FGF2 in neuronal stem cells have provided evidence that FGF2 is required for the production of new dentate granule neurons in the adult brain (Zhao et al. 2007). The blunted neurogenesis resulting from impaired FGF2 signaling in stem cells is associated with both reduced LTP at glutamatergic synapses in the dentate gyrus and impaired memory.

In response to exercise, muscle cells release IGF1 into the blood, and the IGF1 moves across the blood–brain barrier and into the brain. When mice are given IGF1 antibodies intravenously, the ability of exercise to enhance neurogenesis and increase synapse numbers in hippocampal neurons is attenuated (Glasper et al. 2010). Studies in rats have shown that infusion of IGF1 into the cerebral ventricles can enhance performance on learning and memory tests.

Running-wheel exercise may stimulate mitochondrial biogenesis and stress resistance in neurons. Exercise results in a robust increase in BDNF levels in the hippocampus of mice (Cotman and Berchtold 2002). Aiwu Cheng found that BDNF stimulates mitochondrial biogenesis and the formation of new synapses in cerebral cortical neurons (A. Cheng, Wan, et al. 2012). When she used genetic-engineering technology to prevent mitochondrial biogenesis, BDNF was no longer able to stimulate the formation of new synapses. Cheng also found that running-wheel exercise increases the amount of the enzyme SIRT3 in neurons. SIRT3 is located in the mitochondria, where it functions to enhance the mitochondria's ability to cope with the oxidative stress that neurons experience during their normal activity. When mice were treated with a drug that inhibits NMDA receptors, the ability of exercise to increase SIRT3 production was abolished (A. Cheng, Yang, et al. 2016). Therefore, activation of glutamate receptors is required for exercise to bolster mitochondrial stress resistance.

Exercise has also been shown to have antidepressant and anti-anxiety effects in rats, mice, and humans. These effects are believed to be mediated, at least in part, by BDNF (Castren and Kojima 2017). Exercise increases the production of BDNF in brain regions implicated in anxiety disorders and depression, including the hippocampus, amygdala, and prefrontal cortex. Treatments prescribed for patients with these mood disorders—serotonin and norepinephrine reuptake inhibitors, electroconvulsive shock therapy, and ketamine—are known to increase the production of BDNF.

The elevation of BDNF levels with these treatments results from increased activity in glutamatergic neuronal circuits. Animal studies have shown that experimental reduction of BDNF levels in the brain results in depression-like behaviors. Additional evidence suggests that the abilities of antidepressant drugs and exercise to reduce anxiety and depression-like behaviors require BDNF signaling and the transcription factor CREB (Nair and Vaidya 2006).

In addition to enhancing cognition and mood, exercise may protect neurons against traumatic injury and neurodegenerative disorders (Intlekofer and Cotman 2013; Mattson 2012; Patten et al. 2015). Regular aerobic exercise reduces the risk of stroke by lowering blood pressure and reducing atherosclerosis in cerebral arteries. Studies of rats and mice have shown that running-wheel exercise can protect neurons against dysfunction and degeneration in experimental models of stroke and epilepsy. Running-wheel exercise also delays the accumulation of amyloid and associated cognitive impairment in transgenic mouse models of Alzheimer's disease. As described in chapters 7 and 8, the degeneration of neurons in epilepsy, stroke, Alzheimer's disease, Parkinson's disease, and Huntington's disease involves excitotoxicity. By increasing neurons' resistance to excitotoxicity, exercise may thus also provide protection against a wide array of brain disorders.

Numerous studies with cultures of neurons from the hippocampus and cerebral cortex of rats or mice have demonstrated the abilities of BDNF, FGF2, and IGF1 to protect neurons from being damaged and killed by conditions relevant to neurological disorders. BDNF, FGF2, and IGF1 can protect hippocampal neurons against excitotoxicity and glucose deprivation (B. Cheng and Mattson 1992b; B. Cheng and Mattson, 1994). Because aerobic exercise increases the production of BDNF, FGF2, and IGF1, it seems likely that these neurotrophic factors contribute to the ability of exercise to protect neurons in animal models of epilepsy and stroke.

Finally, beyond its beneficial effects on the function and resilience of glutamatergic neurons, exercise can stimulate the proliferation of blood vessels in the brain (Morland et al. 2017). It is well known that regular exercise stimulates angiogenesis (the formation of new blood cells) in skeletal muscles and the heart. This effect of exercise on blood vessels is mediated by a protein called "vascular endothelial cell growth factor," or VEGF. Animal studies have provided evidence that the VEGF produced in muscle cells is released into the blood and thus enters the brain. An increase in VEGF and blood vessel density in the brain would be expected to enable increased provision of nutrients to neurons.

THE CHALLENGE OF GOING WITHOUT

In the mid-1990s, I became familiar with studies showing that the life spans of laboratory rats and mice can be profoundly increased by simply reducing the amount of food they eat each day. Their life span is similarly extended when they are deprived of food every other day. When begun in young adult animals, such daily calorie restriction and intermittent fasting can extend their life span by as much as 80 percent (Goodrick et al. 1982). Because aging is the major risk factor for stroke, Parkinson's, and Alzheimer's diseases, I decided to see whether dietary energy restriction would protect neurons against dysfunction and degeneration in animal models of these disorders. The design of such experiments was simple. Animals were randomly assigned to either an intermittent-fasting group (i.e., with food deprivation every other day) or an ad libitum feeding control group (i.e., with unlimited food available).

During the late 1990s and early 2000s, postdoctoral fellows in my laboratory showed that intermittent fasting can protect neurons against dysfunction and degeneration in a rat model of Huntington's disease, a mouse model of Parkinson's disease, and a transgenic mouse model of Alzheimer's disease.

The first experiments were performed by Anna Bruce-Keller in a neurotoxin model of relevance to epilepsy and Alzheimer's disease. The neurotoxin she used was kainic acid (see chapter 6), which causes seizures and selectively kills hippocampal glutamatergic pyramidal neurons, resulting in profound deficits in spatial learning and memory. This animal model applies mostly to epilepsy but is also relevant to Alzheimer's disease because hippocampal pyramidal neurons are among the most prone to degeneration in Alzheimer's disease. Kainic acid was administered to rats after they had been maintained on the intermittent-fasting or control-feeding regimens for three months. The learning and memory abilities of both groups of rats were then evaluated in a water maze. The rats were then euthanized, and their brains removed and cut into thin slices. The brain slices were stained with a dye that selectively accumulates in neurons, and the numbers of pyramidal neurons in the hippocampus were counted. The results were clear. Whereas rats in the control group had extensive loss of pyramidal neurons and a severe learning and memory deficits, those in the intermittent-fasting group did not (Bruce-Keller, Umberger, et al. 1999).

People with epilepsy often benefit from eating a ketogenic diet, which is a diet high in fat but few or no carbohydrates. The term *ketogenic* refers to the fact that a high-fat diet that lacks carbohydrates results in the fats being converted to ketone molecules in the liver. The ketones beta-hydroxybutyrate and acetoacetate are then released into the blood and serve as a very efficient source of energy for neurons and other cells when glucose levels are low. During fasting, levels of ketones are produced from fats (fatty acids) released from fat cells. In this way, ketones enable people to live for many weeks or even months with no food intake depending on how much fat they have in their body. Ketones are thought to suppress seizures by enhancing the activity of inhibitory GABAergic neurons.

Intermittent fasting may also prevent or reduce seizures by enhancing the activity of inhibitory GABAergic neurons. Yong Liu recorded the amount of inhibition of glutamatergic neurons by GABAergic neurons in the hippocampus of mice that had been maintained on an every-other-day fasting regimen and of control mice that were fed ad libitum. He found that the frequency of release of GABA from the inhibitory neurons was greater in the mice that had been on the intermittent-fasting eating pattern (Y. Liu et al. 2019). In addition to enhancing GABAergic inhibition of glutamatergic neurons, ketogenic diets and fasting may protect neurons against seizures by reducing the production of free radicals in the neurons' mitochondria because more free radicals are produced when neurons use glucose as their energy source than when they use ketones.

Chapter 7 described an animal model of stroke. Using this stroke model, Zaifang Yu found that the amount of brain damage caused by the experimental stroke was reduced and functional outcome improved in rats that had been maintained on an intermittent-fasting feeding regimen compared to the control rats fed ad libitum (Yu and Mattson 1999). In a subsequent study, Thiruma Arumugam provided evidence that the mechanism by which intermittent fasting protects neurons against stroke involves stimulation of the production of neurotrophic factors, heat-shock proteins, and the antioxidant enzyme heme oxygenase 1 (Arumugam et al. 2010).

Anna Bruce-Keller found that intermittent fasting can protect GABAergic medium spiny neurons in the striatum of rats against damage caused by the mitochondrial toxin 3-nitropropionic acid, or 3-NPA (Bruce-Keller, Umberger, et al. 1999). As described in chapter 6, this model is of relevance to Huntington's disease because the GABAergic neurons in the striatum are the first to degenerate in this disease. Bruce-Keller found that intermittent fasting in this animal model significantly reduces damage to the GABAergic neurons and improves control of body movements. Wenzhen Duan found that intermittent fasting also protects dopaminergic neurons in the MPTP mouse model of Parkinson's disease (Duan and Mattson 1999). The toxins used in these models impair the function of mitochondria, resulting in energy deficits in the striatal and dopaminergic neurons. The neurons continue to be stimulated by glutamate but become unable to restore their membrane potential as a result of failure of their ATP-dependent Na^+ and Ca^{2+} pump proteins.

It is likely that intermittent fasting lessens the energy deficits caused by 3-NPA, MPTP, and a stroke. Such cellular energy preservation may occur in several ways. One way is by increasing the production or activity of antioxidant enzymes that remove free radicals and thereby reduce the amount of free radical damage to proteins, membranes, and DNA in mitochondria. In this way, intermittent fasting might protect mitochondria against dysfunction caused by the free radicals produced in the neurons affected in Huntington's disease, Parkinson's disease, and stroke. As evidence, Aiwu Cheng and Yong Liu found that intermittent fasting increases the production of the mitochondrial enzyme SIRT3. In turn, SIRT3 increases the activity of SOD2, which is the most important antioxidant enzyme in the mitochondria of neurons (A. Cheng, Yang, et al. 2016; Y. Liu et al. 2019). Intermittent fasting is also known to stimulate the processes of mitophagy and mitochondrial biogenesis. Chapter 5 described how mitophagy removes unhealthy mitochondria from cells and how mitochondrial biogenesis produces new pristine mitochondria. Thus, intermittent fasting may increase the numbers of healthy, stress-resistant mitochondria in neurons. With more healthy mitochondria, neurons are better able to produce the ATP necessary to protect themselves against glutamate excitotoxicity.

In 2003, Salvador Oddo, Frank LaFerla, and their colleagues at the University of California, Irvine, reported the use of cutting-edge genetic-engineering technologies to produce mice that develop pathologies in their brains and associated cognitive deficits that are very similar to those that occur in Alzheimer's disease (Oddo et al. 2003). They called the transgenic mice "3xTgAD mice" because they express three different mutated human genes associated with Alzheimer's. Two of the mutant genes, those encoding the beta-amyloid (Aβ) precursor protein, or APP, and presenilin-1, cause the accumulation of Aβ plaques in the brains of the mice. The third mutant gene encodes Tau, a protein that accumulates and forms neurofibrillary tangles inside of neurons in Alzheimer's disease and a related disease called "fronto-temporal dementia." As the 3xTgAD mice age, there is a progressive accumulation of Aβ plaques and Tau tangles in them. The plaques and tangles accumulate in large numbers particularly in the hippocampus and connected regions of the cerebral cortex. Learning and memory deficits and impaired hippocampal synaptic plasticity develop in association with amyloid and Tau pathologies.

Veerendra Halagappa tested the hypothesis that intermittent fasting could mitigate the plaque and tangle pathologies as well as the associated cognitive deficits in 3xTgAD mice (Halagappa et al. 2007). When the mice were young (five months old), he randomly assigned them to three different groups: one group fed ad libitum, a second made to fast every other day, and a third fed daily an amount of food that was 40 percent less than the mice fed ad libitum. To establish the cognitive performance of mice without Alzheimer's disease pathologies, the study also included a group of control mice that did not have the mutant human genes and were fed ad libitum. One year later, when the mice were 17 months old, their learning and memory were evaluated in a water maze. Compared to the control mice without the mutant genes who were fed ad libitum, the 3xTgAD mice fed ad libitum performed very poorly in the water maze. It took them much longer to learn the location of the escape platform, and even if they learned its location, they quickly forgot it. In contrast, the 3xTgAD mice in the every-other-day fasting and 40 percent calorie restriction groups performed as well as the mice without Alzheimer's disease pathologies.

Halagappa's study also resulted in a very surprising result. Although intermittent fasting prevented age-related cognitive deficits in the 3xTgAD mice, it did not lessen the amount of amyloid plaques in their brains. The intermittent fasting protected neurons against the adverse effects of the Aβ protein on their structure and function. This is very interesting because studies have shown that some very old people with competent cognitive function nevertheless have very large amounts of amyloid plaques in their brains. There apparently is something about their lifestyle and/or genetic constitution that bolsters their neurons' ability to resist the toxic effects of amyloid. Although it is not known why some people can tolerate extensive accumulation of the Aβ protein in their brains, it is known that being overweight and having insulin resistance increases one's risk for Alzheimer's disease.

Intermittent fasting can protect transgenic mice with Aβ plaques against cognitive impairment. When fed ad libitum, transgenic mice that accumulate amyloid in their brains develop seizures as they age, but not so when they are maintained on an intermittent-fasting feeding schedule. Analyses of synaptic function in the hippocampus demonstrated a deficit in LTP in the mice with Aβ plaques fed ad libitum but no deficit in those on intermittent fasting (Y. Liu et al. 2019). Analyses of the brains of the non-Alzheimer's mice to the Alzheimer's mice revealed that levels of the mitochondrial enzyme SIRT3 were reduced in the hippocampus of the Alzheimer's mice. Moreover, intermittent fasting prevented the reduction in SIRT3 levels.

To determine whether a reduction in SIRT3 levels is sufficient to cause neuronal network hyperexcitability, Yong Liu and colleagues recorded activity of GABAergic neurons in the hippocampus of mice that had been genetically engineered to disable the gene-encoding SIRT3. He found that in these mice intermittent fasting did not increase GABAergic activity. This result demonstrated that mitochondrial SIRT3 was essential for intermittent fasting to reduce neuronal hyperexcitability. The hyperexcitability of neurons in the hippocampus caused by a deficiency of SIRT3 was associated with impaired learning and memory. Moreover, the mice deficient in SIRT3 exhibited increased levels of anxiety. This is interesting because anxiety levels are increased in people with Alzheimer's disease and in mouse models of this disease.

As described in chapter 8, the axons of parasympathetic neurons in the brainstem course through the vagus nerve to the heart and gut. Their activity slows heart rate and stimulates gut motility. In Parkinson's disease, the protein alpha-synuclein accumulates in these neurons, and so they degenerate, resulting in an elevated heart rate and constipation. Transgenic mice that produce a mutant form of alpha-synuclein that causes inherited Parkinson's disease in humans develop progressive deterioration of the motor function, are eventually unable to feed themselves, and have to be euthanized. Examination of their brains reveals progressive accumulation of alpha-synuclein in dopaminergic neurons.

Kathleen Griffioen and Ruiqian Wan performed experiments to determine whether the alpha-synuclein mutant mice develop dysfunction of their brainstem parasympathetic neurons (Griffioen et al. 2013). In these mice, the brainstem parasympathetic neurons accumulate alpha-synuclein, and heart rate is elevated. Alpha-synuclein transgenic mice and mice without the alpha-synuclein mutation were divided into three groups each. One group was fed the normal diet ad libitum; a second was fed a diet with high amounts of fructose and sucrose ad libitum and a third was fed the normal diet and fasted every other day. To record heart rate continuously, transmitters were implanted in all of the mice, with an electrode placed next to the heart. The transmitter sent a signal to a receiver pad placed under each mouse's cage and connected to a computer. Heart rate was recorded during a 12-week period. The heart rate was elevated in alpha-synuclein mutant mice compared to nonmutant mice. The most striking result was that the mice in the intermittent-fasting groups had a reduced heart rate compared to those on the control diet or high-sugar diets. This was true in both the alpha-synuclein mutant mice and the nonmutant mice. Alpha-synuclein mice on the high-sugar diet had an elevated heart rate compared to the alpha-synuclein mice on the control diet. The results of additional experiments showed that alpha-synuclein accumulated in brainstem autonomic neurons of the alpha-synuclein mutant mice, which is associated with a diminished parasympathetic control of heart rate. Intermittent fasting prevented this decline of parasympathetic tone.

In a collaboration with David Mendelowitz, Ruiqian Wan provided evidence that BDNF enhances the function of brainstem parasympathetic neurons and so slows heart rate (Wan et al. 2014). This effect of BDNF results from shifting the relative balance of glutamatergic and GABAergic inputs to the parasympathetic neurons. Both intermittent fasting and aerobic exercise can increase the production of BDNF, which could explain the ability of intermittent fasting and exercise to reduce heart rate. It may also explain the ability of intermittent fasting to prevent dysfunction and degeneration of brainstem cholinergic neurons in alpha-synuclein mutant transgenic mice.

Transgenic mice expressing a mutant human huntingtin gene develop progressive motor dysfunction associated with accumulation of huntingtin protein aggregates in striatal and cerebral cortical neurons. In this case, too, intermittent fasting slows the progression of the brain pathology and motor dysfunction (Duan, Guo, et al. 2003). As in humans with Huntington's disease, the huntingtin mutant mice exhibit reduced levels of BDNF in their striatum and cerebral cortex. Intermittent fasting normalized BDNF levels in the huntingtin mutant mice, suggesting a role for BDNF in counteracting the disease process.

From an evolutionary perspective, intellectual challenges, exercise, and fasting occur simultaneously when animals are hungry and motivated to acquire food. One might therefore expect that the brains of modern-day humans might benefit most from incorporating these three bioenergetic challenges into their lifestyles. However, studies designed to test this hypothesis in humans are lacking. Individual studies typically focus on only one type of challenge and not the others. For example, studies that focus on intellectual challenges may not account for the possibility that more highly educated people tend to get more aerobic exercise and not overeat compared to less-educated people. Well-controlled human studies in which the results of the combination of exercise, intellectual challenges, and intermittent fasting are compared to the results of individual challenges are therefore needed.

But animal studies are beginning to elucidate how multiple environmental challenges interact in ways that affect the structure and function of glutamatergic circuits. Much of this work has focused on the hippocampus. Alexis Stranahan performed experiments in mice aimed at determining whether exercise and daily time-restricted eating with caloric restriction have combinatorial effects on the plasticity of glutamatergic neurons of the hippocampal dentate gyrus (Stranahan, Lee, et al. 2009). The study used both healthy mice and mice with diabetes. The healthy mice and diabetic mice were divided into four groups each. One group was fed ad libitum and was sedentary. A second group was fed ad libitum and had a running wheel in their cage. The third group was sedentary and was maintained on a reduced-calorie diet that they consumed in the same approximately four-hour period daily (daily restricted eating). The fourth group ran daily while on daily restricted eating. After three months, the mice were euthanized and their brains removed. Brain sections on one side of the brain of each mouse were stained with Golgi's stain, and the numbers of dendritic spines on glutamatergic neurons in the hippocampal dentate gyrus were counted. The hippocampus on the other side of the brain was removed and used for measurements of levels of BDNF.

The data revealed that in healthy mice both running and restricted eating resulted in increased numbers of dendritic spines compared to the healthy mice that were sedentary and fed ad libitum. The combination of running and restricted eating resulted in a greater increase in dendritic spine numbers compared to either running or restricted eating alone. Mice with diabetes had reduced numbers of dendritic spines compared to nondiabetic mice regardless of which group they were in. However, running and restricted eating significantly increased dendritic spine numbers in the diabetic mice, and the combination of running and restricted eating further increased synapse numbers. Measurements of levels of BDNF revealed that both running and restricted eating increased levels of this neurotrophic factor in normal and diabetic mice. These findings show that restricted eating and exercise can have additive effects on the structure of glutamatergic hippocampal circuits and that these effects are associated with BDNF production.

Intermittent fasting is known to improve general health in humans (de Cabo and Mattson, 2019; Mattson 2022). It facilitates loss of abdominal fat, increases insulin sensitivity, and improves multiple markers of risk for cardiovascular disease. On the one hand, the effects of intermittent fasting and caloric restriction on the human brain remain to be determined. On the other hand, there is abundant evidence that overeating and obesity adversely affect the human brain. Animal studies have provided insight into how the structure and function of brain neuronal networks can suffer from a "couch potato" lifestyle. A final section of this chapter describes such evidence.

COMPLACENCY ENGENDERS DESTRUCTION

Obesity is a major health problem in industrialized countries. Approximately 33 percent of people in the United States have a body mass index (BMI) higher than 30 and so are considered to have obesity (BMIs between 18.5 and 25 are considered healthy). The prevalence of obesity has increased dramatically since the 1980s, when less than 15 percent of people in the country had obesity. The prevalence of childhood and adolescent obesity increased from 5 percent to 25 percent between 1980 and 2020. Obesity increases the risk for essentially all major causes of death, including cardiovascular disease, stroke, diabetes, cancers, and kidney disease. Being overweight (BMI between 25 and 30) also increases the risk for these diseases, albeit to a lesser degree than obesity.

Several factors have contributed to the obesity epidemic, especially increased consumption of calorie-dense processed foods and reduced levels of physical activity. Consumption of simple sugars such as fructose and saturated fats are thought to be particularly problematic. When a person becomes obese, abnormalities in neuroendocrine systems play a major role in perpetuating the obesity. In particular, in people with obesity, neurons in their hypothalamus have an impaired ability to respond to the "satiety hormone" leptin.

Leptin is produced in fat cells and is released into the blood in response to food consumption, much as insulin is released from pancreatic beta-cells when glucose levels rise. Neurons in the hypothalamus at the base of the brain respond to leptin, resulting in the activation of neuronal networks that inhibit food intake and so prevent overeating. In people who are obese, the neurons in the hypothalamus do not respond well to leptin, and so such individuals continue to feel hungry even though they are overeating. This condition is called "leptin resistance" and is analogous to insulin resistance, wherein liver, muscle, and other cells in the body become relatively unresponsive to insulin, and so blood sugar levels remain abnormally high. The good news is that just as exercise and intermittent fasting can prevent and reverse insulin resistance, they can also prevent and reverse leptin resistance. (Mattson 2022, 184)

Whereas exercise and moderation in calorie intake have beneficial effects on glutamatergic neuronal networks, sedentary and overindulgent lifestyles have detrimental effects. Data from epidemiological studies have shown that, on average, children and adults with obesity and/or diabetes perform relatively poorly on learning and memory tests compared to their non-obese and metabolically healthy peers. Po Yau, Antonio Convit, and colleagues at New York University tested the cognitive abilities and performed brain-imaging analyses of 62 adolescents who were overweight and had insulin resistance and 49 control adolescents who were not overweight and did not have insulin resistance (Yau et al. 2012). They found that compared to latter group, those who were metabolically unhealthy performed significantly more poorly on spelling, arithmetic, metabolic flexibility, and attention. Analyses of MRI images revealed that the metabolically unhealthy adolescents had smaller hippocampal volumes. They also had enlarged cerebral ventricles, suggesting a general reduction in gray matter in the cerebral cortex.

In one study, MRI images of the brains of 420 people were acquired when they were 60–64 years old and again eight years later. Analyses of the images revealed a negative correlation between BMI and size of the hippocampus (Cherbuin et al. 2015). As BMI increased, the size of the hippocampus decreased. The study also showed that people with higher BMIs showed a greater atrophy of the hippocampus during the eight-year period.

Working at the National Institute of Neuroscience in Japan, Shinsuke Hidese and colleagues showed that among people with depression, those with obesity have much greater shrinkage of the hippocampus compared to those without obesity (Hidese et al. 2018). Cognitive testing revealed that working memory, verbal memory, attention, and executive function were reduced more in depressed people with obesity compared to depressed people without obesity. The good news is that there is evidence that the reductions in the size of the hippocampus that occurs in depression and obesity can be reversed with recovery from the depression and reduction in body weight.

How might obesity and diabetes adversely affect glutamatergic neuronal networks in ways that diminish cognitive abilities and increase the risks for anxiety disorders, depression, Alzheimer's disease, and stroke? Numerous studies in rats and mice have shown that obesity and diabetes can impair learning and memory (Mattson 2019). Such studies include models in which obesity and diabetes are caused by diets high in glucose, fructose, and/or saturated fats. Mice with genetic mutations that cause obesity and diabetes have also been used, including mice in which the leptin gene is disabled. Here are a few examples of such studies.

Shuai Hao, Alexis Stranahan, and coworkers at Augusta University maintained mice on either a high-fat diet that causes obesity or the usual lower-fat diet for three months (Hao et al. 2016). They then evaluated the learning and memory of the mice in maze and novel-object recognition tests. In addition, LTP at glutamatergic synapses on dentate gyrus granule neurons and synapse numbers were measured. The mice with obesity exhibited poorer learning and memory, impaired LTP, and reduced numbers of synapses. When they were put on the lower-fat diet for two months, they lost body weight, and the adverse effects of obesity on hippocampal glutamatergic circuits were reversed. Additional experiments provided evidence that the high-fat diet caused the activation of microglia and that those microglia stripped glutamatergic synapses from dentate granule neuron dendrites.

The mechanisms by which lack of exercise and excessive energy intake adversely affect the brain involve disengagement of processes by which exercise and energy restriction enhance brain health. In essence, without being regularly challenged, neurons become complacent. The evidence for this "complacency hypothesis" is considerable (Mattson, Moehl, et al. 2018). In animal models of obesity and diabetes, the production of BDNF is reduced, neuronal network hyperexcitability occurs, and GABAergic tone is reduced. In addition, obesity and diabetes compromise mitochondrial function, autophagy, DNA repair, and antioxidant defenses. These cellular and molecular consequences of obesity and diabetes likely contribute to the brain's increased vulnerability to cognitive impairment and Alzheimer's disease. As evidence, it has been demonstrated that diet-induced obesity accelerates brain neuropathology and cognitive impairment in mouse models of Alzheimer's disease.

Another adverse consequence of obesity and diabetes on the brain concerns insulin. Studies of animal models suggest that neurons in the brain can develop insulin resistance. Claudia Grillo, Lawrence Reagan, and their coworkers at Columbia University used molecular genetic methods to selectively deplete the insulin receptor in the hippocampal neurons of rats. The rats whose hippocampal neurons could not respond to insulin exhibited impaired LTP and cognition (Grillo et al. 2015). This result suggests that neurons' resistance to insulin could play a role in the adverse effects of obesity and diabetes on glutamatergic neuronal circuits. Data from human studies support the notion that neurons' ability to respond to insulin is impaired in diabetes. For example, a study by Dimitrios Kapogiannis and colleagues at the US National Institute on Aging showed that activation of the insulin signaling is impaired in neuron-derived membrane vesicles isolated from the blood of people with diabetes and preclinical Alzheimer's disease compared to people without these conditions (Kapogiannis et al. 2015). These findings suggest that neurons' impaired ability to respond to insulin may contribute to the adverse effects of obesity and diabetes on cognition.

Inflammation throughout the body is a common result of the excessive accumulation of abdominal fat and alterations in the gut bacterial flora that occur in obesity. Evidence also suggests that the brain is not spared from such inflammation. The hypothalamus was the first brain region shown to be affected by inflammation caused by obesity. This "neuroinflammation" is characterized by the activation of microglia and increased production of inflammatory cytokines. Evidence suggests that this local inflammation in the hypothalamus contributes to the hypothalamic neurons' impaired ability to respond to the satiety hormone leptin. Hippocampal neuroinflammation caused by obesity also likely contributes to impaired cognition (Guillemot-Legris and Muccioli 2017). Increased inflammatory cytokine levels and microglial activation have been documented in animal models of obesity. Moreover, inflammation even in the absence of obesity in animal models—typically induced by administering a bacterial molecule called "lipopolysaccharide"—is sufficient to impair hippocampal synaptic plasticity and cognition.

Neuroinflammation occurs in several major brain disorders, including Alzheimer's disease, Parkinson's disease, and major depression. It is thought that the low-grade neuroinflammation that occurs in obesity accelerates the disease process in these disorders. Therefore, in addition to the disengagement of the mechanisms by which exercise and energy restriction protect the brain, neuroinflammation may contribute to poor brain health in people with obesity. If someone with obesity loses weight by reducing their energy intake or exercising, their cognition improves (A. Martin et al. 2018).

Obesity and diabetes often result in the elevation of cortisol levels owing to hyperactivation of the neurons in the hypothalamus that produce the adrenocorticotropic hormone. Chapter 9 described how chronic psychological stress elevates cortisol levels and adversely affects glutamatergic neuronal networks. Such stress is clearly bad stress. Studies of the hippocampus showed that chronic psychological stress results in the degeneration of synapses, impaired neurogenesis, and cognitive deficits. In contrast, caloric restriction and intermittent fasting promote the formation of new synapses, stimulate neurogenesis, and enhance cognition.

Excessive cortisol may also contribute to the adverse effects of obesity on the structure and function of glutamatergic circuits. So how can it be that caloric restriction and intermittent fasting elevate cortisol levels but are good for the brain?

When Jaewon Lee was a graduate student in my laboratory, he performed an experiment that helped solve the "cortisol paradox." We knew that glutamatergic neurons have two different receptors for cortisol: the glucocorticoid receptor and the mineralocorticoid receptor. Both receptors are located in the cytoplasm of the neurons. When cortisol enters a glutamatergic neuron and binds to a receptor, the receptor moves into the nucleus. The glucocorticoid and mineralocorticoid receptors are transcription factors that turn certain genes off or on. Previous research demonstrated that chronic activation of the glucocorticoid receptor is responsible for the adverse effects of "bad stress" on glutamatergic neurons. It had also been shown that chronic uncontrollable stress causes a decrease in the levels of the mineralocorticoid receptor in hippocampal neurons. Given that glucocorticoid seemed to mediate the adverse effects of bad stress on the brain, we asked whether mineralocorticoid receptors might contribute to the beneficial effects of the good stress of intermittent fasting.

Lee's experiment was relatively simple. Rats were divided into two groups. The rats in one group were on every-other-day fasting feeding schedule, and the rats in the other group were fed ad libitum. After three months, the rats were euthanized, and their brains were removed and cut into thin slices. A method called "in situ autoradiography" was used to measure the relative amounts of the messenger RNAs encoding the glucocorticoid and mineralocorticoid receptors in hippocampal neurons. The results showed that intermittent fasting decreased the level of the glucocorticoid receptor messenger RNA in hippocampal pyramidal and dentate granule neurons but maintained levels of mineralocorticoid receptor messenger RNA (Lee, Herman, and Mattson 2000). Additional analyses demonstrated that glucocorticoid receptor protein levels were also decreased in response to intermittent fasting. These results provided evidence that although cortisol levels are elevated during intermittent fasting, the way in which glutamatergic hippocampal neurons respond to the cortisol is different than how they respond to chronic psychological stress.

Lee's findings begged the question of whether activation of the mineralocorticoid receptor can counteract the adverse effects of "bad stresses" on hippocampal neuronal networks. To answer this question, Alexis Stranahan used mice genetically engineered so that they do not have functional leptin receptors and so eat excessively and become obese and diabetic. Stranahan found that these diabetic mice have reduced numbers of synapses and impaired LTP at hippocampal glutamatergic synapses, both of which are associated with cognitive deficits (Stranahan, Arumugam, et al. 2008). She also found that the diabetic mice have elevated levels of corticosterone—the mouse equivalent of cortisol—and that lowering corticosterone levels by removing the adrenal glands lessens the adverse effects of the diabetes on the hippocampal circuits.

Having established an important role for corticosterone in the adverse effects of diabetes on hippocampal synapses, Stranahan next asked whether mineralocorticoid receptor activation might protect the synapses against diabetes. To do this, she used a different model of diabetes in rats, in which the animals produce very little insulin and so have very high levels of blood glucose. She then prepared hippocampal slices and performed electrophysiological recordings from dentate granule neurons. LTP at glutamatergic synapses in the dentate gyrus was impaired in the diabetic rats but not in the nondiabetic rats. She treated the hippocampal slices with aldosterone, a hormone that activates the mineralocorticoid receptor but not the glucocorticoid receptor. Aldosterone restored LTP (Stranahan, Arumugam, et al. 2010). When considered in light of Jaewon Lee's findings, Stranahan's data are consistent with the possibility that if the mineralocorticoid receptor is activated, the elevation of cortisol levels that occurs during intermittent fasting has beneficial effects on hippocampal glutamatergic neurons.

In the United States, the prevalence of obesity and associated diseases—including diabetes, cardiovascular disease, stroke, and many types of cancer—is greatest in the southern states of Alabama, Mississippi, Louisiana, Arkansas, Georgia, and Texas. Childhood obesity is also greatest in these states. Children in these states have poorer academic performance and are less likely to go to college. Tests of cognitive function show that obesity is associated with poorer working memory, executive function, and verbal memory (Mattson 2022). Sociologists and epidemiologists have pointed to low socioeconomic status and poorer schools as key factors in these southern children's poor academic performance. However, in light of clear evidence that obesity can compromise cognition, it seems likely that the children's poor metabolic health also contributes to their poor academic outcomes. If so, then their academic performance could be improved by including exercise programs in their schools, improving their diets, and having them adopt an intermittent-fasting eating pattern. To be successful, such efforts must include education of parents on the connection between the children's metabolic health and their academic performance.

SUMMARY

1. Glutamate is an evolutionarily ancient neurotransmitter and mediates the reflex responses of simple organisms to environmental stimuli. The conservation of glutamatergic neurons throughout evolution attests to their fundamental importance.
2. Glutamate plays a major role in the establishment of neuronal networks during brain development.
3. Of the approximately 90 billion neurons in the human brain, more than 90 percent deploy glutamate as their neurotransmitter. Dopaminergic, serotonergic, noradrenergic, and cholinergic neurons are much fewer in number and are confined to small brain regions under the cerebral cortex.
4. Glutamate is the master and commander even of neurons that deploy neurotransmitters other than glutamate.
5. Glutamate is the only neurotransmitter that causes robust excitation of the neurons upon which it acts. Without glutamatergic neurotransmission, the brain would completely shut down. In contrast, if other neurotransmitters are disabled, the brain still functions, albeit not in its normal way.
6. Neuronal circuits throughout the cerebral cortex, hippocampus, and cerebellum consist entirely of excitatory glutamatergic neurons and inhibitory GABAergic neurons. Glutamatergic neurons have long axons that project within and between brain regions, whereas GABAergic neurons have relatively short axons that are confined to local circuits.

7. The GABA in the inhibitory neurons is produced from glutamate. There would be no GABA without glutamate.

8. Glutamate is responsible for the neuroplasticity that occurs in the brain in response to environmental challenges, such as acquiring food, learning how to play a musical instrument, and dealing with a stressful life event.

9. Evidence suggests that changes at glutamatergic synapses are largely responsible for learning and memory and the advanced capabilities of the human brain, including imagination, creativity, and language.

10. Glutamatergic neurotransmission plays a preeminent role in brain energy metabolism. Indeed, fMRI images of the brain reflect the relative activities of glutamatergic neurons.

11. Hyperexcitability of glutamatergic neurons contributes to neuronal degeneration in epilepsy, stroke, and traumatic brain injuries. In addition, excitotoxicity is also involved, albeit more insidiously, in the demise of neurons in Alzheimer's disease, Parkinson's disease, ALS, and Huntington's disease. Moreover, dysregulation of glutamatergic neuronal networks is implicated in psychiatric disorders such as schizophrenia, chronic anxiety, and depression. Dysregulation of glutamatergic signaling is also implicated in some developmental brain disorders—most notably autism.

EPILOGUE

The information provided in this book supports the conclusion that glutamate is the preeminent intercellular signaling molecule that controls the formation, cellular architecture, and function of the brain. Moreover, aberrancies in glutamatergic neuronal networks are involved in all of the major brain disorders that currently plague humankind. This epilogue summarizes the evidence supporting these conclusions.

Very early in the evolution of life on Earth, glutamate functioned as an intercellular signal controlling the growth of multicellular organisms. In slime mold, glutamate is a signal that controls the transition from single-cell amoeba to the multicellular spore-producing organism. Mosses have two glutamate receptors that are critical for reproduction and embryonic development. The growth and branching patterns of the roots and stems of plants are determined in part by glutamate signaling between the cells of those structures. As in neuronal networks, these effects of glutamate on the cellular architecture of plants result from Ca^{2+} influx through membrane channels. Responses of plants to stress, such as leaf damage caused by insects, are also mediated by glutamate and Ca^{2+}. Remarkably, plants have more genes that encode glutamate receptors than do mammals, but the precise roles of the individual receptors in the lives of plants remain to be determined. In insects, glutamate is the neurotransmitter deployed at the neuromuscular synapse and so is essential for locomotion and reflex responses. Studies of insects, roundworms, and mammals have demonstrated evolutionarily conserved roles for glutamate in learning and memory.

In mammals, including humans, more than 90 percent of the neurons in the brain deploy glutamate as a neurotransmitter. Information flow through neuronal circuits within and between all brain regions is conveyed by glutamatergic neurons. Therefore, glutamatergic neurotransmission controls all brain functions, including learning and memory, emotions, creativity, imagination, and decision-making. All other neurotransmitters—GABA, dopamine, serotonin, norepinephrine, and acetylcholine—can exert their effects on behavior only by modulating the activity of glutamatergic neurons.

A remarkable revelation emanating from studies of brain development is that glutamate controls the formation of neuron networks throughout the brain. By causing a highly localized influx of Ca^{2+} in a growing dendrite, glutamate initiates the formation of a synapse. The Ca^{2+} signal causes the activation of genes encoding neurotrophic factors such as BDNF. The neurotrophic factors are released and activate receptors on both the neuron from which the factors are produced and the presynaptic neurons in ways that enable the persistence of the synapse. The neurotrophic factors also promote the neurons' survival and growth. In this way, glutamate determines which neurons form connections and survive and which neurons die by the process of apoptosis. Thus, by causing localized increases in intracellular Ca^{2+} levels and neurotrophic factor production, glutamate determines the number of neurons in the brain, their architecture, and their functional connectivity.

Studies of the hippocampus have conclusively shown that glutamate is the most important neurotransmitter for learning and memory. Learning and memory results from the activation of glutamatergic synapses, Ca^{2+} influx through NMDA receptors, and both rapid and delayed molecular and structural modifications of the active synapses. These memory-encoding changes include the rapid insertion of AMPA receptors into the postsynaptic membrane, activation of the transcription factor CREB, and an increase in the size of the potentiated synapse. The activation of glutamate receptors increases the production of BDNF, which elicits multiple growth responses in both the presynaptic and postsynaptic neurons. These responses include growth, mitochondrial biogenesis, and the formation of new synapses. The postsynaptic Ca^{2+} also triggers the production and release of nitric oxide, which diffuses to adjacent presynaptic axon terminals, where it may elicit changes that result in enhanced glutamate release.

Although major progress has been made in understanding the cellular and molecular mechanisms responsible for learning and memory, the physical nature of a memory remains elusive. The results of recent optogenetic studies have prompted the notion of "engram cells," which are single neurons or a small group of neurons that store an individual memory. But this conceptualization of memories seems overly simplistic because it implies that the brain's capacity is limited by the number of neurons it contains. Individual neurons involved in encoding memories receive hundreds or even thousands of glutamatergic inputs, and evidence suggests that the simultaneous activation of more than one of those synapses is necessary for encoding a memory. Moreover, there is redundancy in the location of stored memories, as demonstrated by the results of brain lesion studies showing that individual memories can be stored in multiple brain regions.

Neuronal networks throughout the brain are active 24/7, and their activity consumes relatively large amounts of energy. Most of the brain's neurons and synapses are glutamatergic, so most of the brain's energy consumption is driven by the activity of glutamatergic neurons. Evidence presented in chapter 4 suggests that the neuroarchitecture of the brain is established in part by processes designed to maximize individual neurons' access to energy. The latter concept is bolstered by comparisons with the well-established mechanisms by which plant branches and roots grow in fractal-like patterns that maximize their access to sunlight and water, respectively. The brain's energy efficiency is enabled by several mechanisms, including activity-dependent mitochondrial biogenesis, the positioning of mitochondria at sites of high energy demand (glutamatergic synapses), the shuttling of energy substrates from astrocytes to neurons, and neurons' ability to utilize ketones as an energy source and as signaling molecules during extended periods of food deprivation.

Glutamate's preeminent role as an orchestrator of the intricate structure and functional capabilities of the brain's neuronal networks begs the question of the consequences of aberrant glutamatergic neurotransmission. Research spanning more than five decades has led to a resounding answer to this question: unconstrained excessive activation of glutamate receptors is responsible for the degeneration and death of neurons in a wide range of neurological disorders. Such excitotoxicity manifests most dramatically in epileptic seizures caused by exposure to environmental toxins or arising idiopathically. Many neurons also die by excitotoxicity in people who suffer a stroke or traumatic brain injury. Evidence from studies of human patients and animal models has shown that neuronal network hyperexcitability occurs insidiously in Alzheimer's disease, Parkinson's disease, Huntington's disease, and ALS. In the latter disorders, aging and disease-specific factors render certain populations of neurons vulnerable to excitotoxicity. In the most common cases of Alzheimer's and Parkinson's diseases, age-related impairments of mitochondrial function, autophagy, and antioxidant defenses may result in an excitatory imbalance that triggers the amyloid, Tau, or alpha-synuclein pathologies that characterize these diseases. In early-onset inherited forms of neurodegenerative disease, the protein encoded by the mutant gene (such as APP, presenilin 1, alpha-synuclein, Parkin, huntingtin, etc.) disturbs neuronal systems that protect neurons against excitotoxicity by mechanisms described in chapter 8.

Psychiatric disorders—including depression, anxiety disorders, PTSD, schizophrenia, and autism spectrum disorders—affect millions of people throughout the world. Historically, drugs that lessen the symptoms of these disorders target synapses that deploy neurotransmitters other than glutamate—serotonin and norepinephrine in depression, GABA in anxiety disorders, dopamine in schizophrenia and autism. However, oft unappreciated is the fact that the vast majority of synapses for these other neurotransmitters are located on glutamatergic neurons. Thus, changes in the activity of glutamatergic circuits are responsible for the symptoms of psychiatric disorders and the lessening of the symptoms by the prescribed drugs. In psychiatric disorders, structural changes in neurons have been documented in one or more brain regions. These changes include a reduction in synapses numbers and dendrite regression in glutamatergic neurons in the hippocampus in depression, increased synapse numbers in the amygdala in anxiety disorders, and exuberant growth of neurons in the frontal cortex during embryonic brain development, resulting in the neuronal network hyperexcitability of autism spectrum disorders. In the case of depression, the shrinkage of the hippocampal neurons is thought to result from a deficit in BDNF production and is reversed with antidepressant treatments that stimulate BDNF production. Studies with the NMDA receptor antagonist ketamine have highlighted major roles for changes in glutamatergic neurotransmission in both the genesis and treatment of psychiatric disorders.

Psychoactive drugs elicit their effects on the brain by two general mechanisms: one is the activation or inhibition of receptors on the dendrites or presynaptic terminals of glutamatergic neurons, and the other is the inhibition of neurotransmitter transporters on the presynaptic terminals of monoaminergic neurons that innervate glutamatergic neurons. In these ways, addictive drugs such as opioids, alcohol, amphetamines, and nicotine increase the activity of dopaminergic neurons in the VTA, which in turn alters the activity of neurons in the nucleus accumbens, prefrontal cortex, amygdala, and hypothalamus. Chapter 10 described the roles of alterations of glutamatergic neurons in addictive drug binging, craving, and withdrawal. A very exciting development in research on psychoactive drugs concerns the remarkable beneficial effects of psychedelics on mood. For example, research with the psilocybin of "magic mushrooms" suggests that it can be effective in treating depression, PTSD, and drug addiction. Such psychedelic drugs are not addictive, are generally safe, and exert long-lasting beneficial effects on mood. Although ketamine and phencyclidine can have effects on mood similar to the effects produced by psychedelics, they are less safe, with a potential for overdose and the triggering of psychosis.

Despite the involvement of aberrant glutamatergic neuronal network activities in many brain disorders, progress in developing treatments that target glutamatergic neurotransmission has been limited. The fact that tightly regulated glutamatergic neurotransmission is required for essentially all functions of the brain presents a problem for drug development. Drugs that inhibit glutamatergic synapses can impair cognition and sensory–motor function; drugs that activate the synapses can cause excitotoxicity. This explains why the US Food and Drug Administration has approved only three drugs that target glutamatergic neurotransmission: ketamine for depression, memantine for Alzheimer's disease, and riluzole for ALS. Although ketamine is proving to be very effective for depression, memantine and riluzole have only modest benefits for Alzheimer's and ALS patients, respectively. This also explains why drugs that affect glutamatergic neurotransmission indirectly and subtly via actions on other neurotransmitter systems can have benefits with tolerable side effects.

This book has described how glutamatergic neurons are fundamental to the brain's neuroarchitecture, functional repertoire, and susceptibility to disease. This begs the question whether there are practical means of optimizing the structure, function, and resilience of the brain's neuronal networks. As described in chapter 11, the answer to this question is yes. This answer is firmly based on the environmental challenges that drove brain evolution, in particular food scarcity. The brains and bodies of all animals, including humans, evolved to function at a high level in a food-deprived state. It is perhaps not surprising, therefore, that animal studies have clearly shown that fasting, physical exercise, and intellectual challenges can improve cognition, mood, and resilience during aging. The underlying mechanisms involve increased activity of glutamatergic circuits throughout the brain, including those engaged in motivation, cognition, decision-making, sensory–motor processing, and neuroendocrine processes. As a result of this increased activity, BDNF production increases, mitochondrial biogenesis occurs, the size and number of synapses increase, and hippocampal neurogenesis occurs. In addition, during fasting and extended exercise, ketones become the main energy source for neurons and affect gene expression in ways that enhance neuronal plasticity and stress resistance.

An overarching feature of science is that new discoveries lead to new questions. It is therefore appropriate to end this book by posing several outstanding questions regarding glutamate's roles as a sculptor and destroyer of the brain's cellular architecture and functions.

Based on the current understanding of the cellular and molecular mechanisms mediating cognitive processes, it seems very likely that increased numbers of glutamatergic neurons and synapses are largely responsible for the human brain's superior functional capabilities. This seems remarkable in light of the fact that the DNA sequences of genes in humans and in our closest animal cousins, chimpanzees, are 99 percent identical. What is the genetic basis for selection of brains with increased numbers of glutamatergic neurons during the evolution of animals? Do only a few genetic mutations account for the expansion of glutamatergic circuits during brain evolution? If so, what are the mechanisms by which the mutations increase the numbers of glutamatergic neurons and synapses? Is there a role for differences in DNA methylation or other such epigenetic mechanisms that affect gene expression in brain evolution?

How will human brains change as evolution proceeds? There is evidence from skull-size measurements that the overall size of the human brain has decreased from times before the Agricultural Revolution to the present day. Which neuronal circuits previously essential for survival in hunter-gatherer environments will be repurposed for the increasingly specialized occupations of modern and future societies? Will some circuits be lost, while others expand?

How are sequences of images and sounds stored in the brain, and how are they recalled? One general approach toward answering these questions is by analogy to sequences of images and sounds stored on and recalled from computer memory chips. In computers, memories consist of sequences of 0s and 1s. The brain's binary code could be 0 = glutamatergic synapse off, 1 = glutamatergic synapse on. Layered on this binary code may be analog mechanisms, including other neurotransmitters' modulatory effects. How does all this work at the levels of both individual cells and entire networks?

How do different signaling systems interact to establish the brain's neuronal networks during brain development? With few exceptions, the individual molecules—neurotrophic factors, cell adhesion molecules, neurotransmitters, and so on—have been studied in isolation. But the signaling mechanisms of these different morphogenic molecules presumably interact in a highly orchestrated manner. There has been progress in elucidating how glutamate and BDNF signaling pathways interact in the process of synaptogenesis, but even this progress falls far short of a full understanding of the mechanisms that sculpt the brain during development.

The calcium ion is the intracellular messenger that mediates the effects of activation of glutamate receptors on neuronal form and function. Although considerable progress has been made in identifying Ca^{2+}-dependent kinases and transcription factors involved in synaptic plasticity, a full understanding of calcium's actions within a neuron is lacking. Open questions include: How are local molecular changes initiated by Ca^{2+} in a dendritic spine coordinated with gene regulation in the nucleus? How does intracellular Ca^{2+} affect the interactions of a growth cone with the adjacent neurons, astrocytes, or extra-cellular matrix molecules? And how are transient Ca^{2+} signals converted into enduring memories?

What explains the increased vulnerability of the brain to neuronal network hyperexcitability and excitotoxicity during aging? Pursuing the answers to this question in particular will be important for developing effective means of preventing and treating Alzheimer's disease and Parkinson's disease.

What contributions does glutamatergic neurotransmission make to the beneficial effects of intellectual challenges, physical exercise, and intermittent fasting on cognition, mood, and stress resistance? In contrast, what are the molecular and cellular mechanisms by which a "couch potato" lifestyle adversely affects glutamatergic neurons? What neuroarchitectural changes occur in response to healthy or unhealthy lifestyles? Which brain regions are affected and in what ways?

Can pharmacological interventions be developed to improve the function and resilience of glutamatergic circuits but with negligible side effects?

These questions are just a few of the innumerable puzzles arising from the research on the neurotransmitter glutamate. I expect that you have already thought of many more.

References

Abbott, A. E., A. Nair, C. L. Keown, M. Datko, A. Jahedi, I. Fishman, and R. A. Muller. 2016. "Patterns of atypical functional connectivity and behavioral links in autism differ between default, salience, and executive networks." *Cerebral Cortex* 26:4034–4045.

Abdelfattah, A. S., T. Kawashima, A. Singh, O. Novak, H. Liu, Y. Shuai, et al. 2019. "Bright and photostable chemigenetic indicators for extended in vivo voltage imaging." *Science* 365:699–704.

Abe, K., and J. Kimura. 1996. "The possible role of hydrogen sulfide as an endogenous neuromodulator." *Journal of Neuroscience* 16:1066–1071.

Abraham, W. C., and W. Tate. 1997. "Metaplasticity: A new vista across the field of synaptic plasticity." *Progress in Neurobiology* 52:303–323.

Alasmari, F., S. Goodwani, R. E. McCullumsmith, and Y. Sari. 2018. "Role of glutamatergic system and mesocorticolimbic circuits in alcohol dependence." *Progress in Neurobiology* 171: 32–49.

Albensi, B. C., and M. P. Mattson. 2000. "Evidence for the involvement of TNF and NF-kappaB in hippocampal synaptic plasticity." *Synapse* 35:151–159.

Amellem, I., G. Yovianto, H. T. Chong, R. R. Nair, V. Cnops, A. Thanawalla, and A. Tashiro. 2021. "Role of NMDA receptors in adult neurogenesis and normal development of the dentate gyrus." *eNeuro* 8. doi: 10.1523/ENEURO.0566-20.2021.

Arancibia, S., M. Silhol, F. Moulier, J. Meffre, I. Hollinger, T. Maurice, et al. 2008. "Protective effect of BDNF against beta-amyloid induced neurotoxicity in vitro and in vivo in rats." *Neurobiology of Disease* 31:316–326.

Arumugam, T. V., T. M. Phillips, A. Cheng, C. H. Morrell, M. P. Mattson, and R. Wan. 2010. "Age and energy intake interact to modify cell stress pathways and stroke outcome." *Annals of Neurology* 67:41–52.

Ashdown-Franks, G., J. Firth, R. Carney, A. F. Carvalho, M. Hallgren, A. Koyanagi, et al. 2020. "Exercise as medicine for mental and substance use disorders: A meta-review of the benefits for neuropsychiatric and cognitive outcomes." *Sports Medicine* 50:151–170.

Athauda, D., K. Maclagan, S. S. Skene, M. Bajwa-Joseph, D. Letchford, K. Chowdhury, et al. 2017. "Exenatide once weekly versus placebo in Parkinson's disease: A randomised, double-blind, placebo-controlled trial." *Lancet* 390:1664–1675.

Augustinack, J. C., A. J. van der Kouwe, D. H. Salat, T. Benner, A. A., Stevens, J. Annese, et al. 2014. "H.M.'s contributions to neuroscience: A review and autopsy studies." *Hippocampus* 24:1267–1286.

Bakker, A., G. L. Krauss, M. S. Albert, C. L. Speck, L. R. Jones, C. E. Stark, et al. 2012. "Reduction of hippocampal hyperactivity improves cognition in amnestic mild cognitive impairment." *Neuron* 74:467–474.

Banack, S. A., T. A. Caller, and E. W. Stommel. 2010. "The cyanobacteria derived toxin beta-N-methylamino-L-alanine and amyotrophic lateral sclerosis." *Toxins* 2:2837–2850.

Barger, S. W., D. Horster, K. Furukawa, Y. Goodman, J. Krieglstein, and M. P. Mattson. 1995. "Tumor necrosis factors alpha and beta protect neurons against amyloid beta-peptide toxicity: Evidence for involvement of a kappa B-binding factor and attenuation of peroxide and Ca2$^+$ accumulation." *Proceedings of the National Academy of Sciences USA* 92:9328–9332.

Barger S. W., and M. P. Mattson. 1996. "Induction of neuroprotective κ B-dependent transcription by secreted forms of the Alzheimer's β-amyloid precursor." *Molecular Brain Research* 40:116–126.

Barrett, F. S., M. K. Doss, N. D. Sepeda, J. J. Pekar, and R. R. Griffiths. 2020. "Emotions and brain function are altered up to one month after a single high dose of psilocybin." *Science Reports* 10. doi: 10.1038/s41598-020-59282-y.

Bazzigaluppi, P., E. M. Lake, T. L. Beckett, M. M. Koletar, I. Weisspapir, S.Heinen, et al. 2018. "Imaging the effects of β-hydroxybutyrate on peri-infarct neurovascular function and metabolism." *Stroke* 49:2173–2181.

Beal, M. F., R. T. Matthews, A. Tieleman, and C. W. Shults. 1998. "Coenzyme Q10 attenuates the 1-methyl-4-phenyl-1,2,3,tetrahydropyridine (MPTP) induced loss of striatal dopamine and dopaminergic axons in aged mice." *Brain Research* 783:109–114.

Ben-Ari, Y., I. Khalilov, K. T. Kahle, and E. Cherubini. 2012. "The GABA excitatory/inhibitory shift in brain maturation and neurological disorders." *Neuroscientist* 18:467–486.

Bergles, D. E., J. D. Roberts, P. Somogyi, and C. E. Jahr. 2000. "Glutamatergic synapses on oligodendrocyte precursor cells in the hippocampus." *Nature* 405:187–191.

Bezprozvanny, I., and M. P. Mattson. 2008. "Neuronal calcium mishandling and the pathogenesis of Alzheimer's disease." *Trends in Neurosciences* 31:454–463.

Bhattacharya, A., H. Kaphzan, A. C. Alvarwz-Dieppa, J. P. Murphy, P. Pierre, and E. Klann. 2012. "Genetic removal of p70 S6 kinase 1 corrects molecular, synaptic, and behavioral phenotypes in fragile X syndrome mice." *Neuron* 76:325–337.

Bielak, A. A. 2010. "How can we not 'lose it' if we still don't understand how to 'use it'? Unanswered questions about the influence of activity participation on cognitive performance in older age—a mini-review." *Gerontology* 56:507–519.

Biscoe, T. J., R. H. Evans, A. A. Francis, M. R. Martin, J. C. Watkins, J. Davies, et al. 1977. "D-alpha-Aminoadipate as a selective antagonist of amino acid–induced and synaptic excitation of mammalian spinal neurones." *Nature* 270:743–745.

Bishop, M. W., S. Chakraborty, G. A. Matthews, A. Dougalis, N. W. Wood, R. Festenstein, et al. 2010. "Hyperexcitable substantia nigra dopamine neurons in PINK1- and HtrA2/Omi-deficient mice." *Journal of Neurophysiology* 104:3009–3020.

Bjork, J. M., and J. M. Gilman. 2014. "The effects of acute alcohol administration on the human brain: Insights from neuroimaging." *Neuropharmacology* 84:101–110.

Bliss, T. V., G. L. Collingride, B. K. Kaang, and M. Zhuo. 2016. "Synaptic plasticity in the anterior cingulate cortex in acute and chronic pain." *Nature Reviews Neuroscience* 17:485–496.

Bogaert, E., C. d'Ydewalle, and L. Van Den Bosch. 2010. "Amyotrophic lateral sclerosis and excitotoxicity: From pathological mechanism to therapeutic target." *Neurological Disorder Drug Targets* 9:297–304.

Boulter, J., M. Hollmann, A. O'Shea-Greenfield, M. Hartley, E. Deneris, C. Maron, et al. 1990. "Molecular cloning and functional expression of glutamate receptor subunit genes." *Science* 249:1033–1037.

Braak, H., R. A. de Vos, J. Bohl, and K. Del Tredici. 2006. "Gastric alpha-synuclein immunoreactive inclusions in Meissner's and Auerbach's plexuses in cases staged for Parkinson's disease-related brain pathology." *Neuroscience Letters* 396:67–72.

Braat, S., and R. F. Kooy. 2015. "The GABAA receptor as a therapeutic target for neurodevelopmental disorders." *Neuron* 86:1119–1130.

Bredt, D. S., and S. H. Snyder. 1990. "Isolation of nitric oxide synthetase, a calmodulin-requiring enzyme." *Proceedings of the National Academy of Sciences USA* 87:682–685.

Brouillet, E., and M. F. Beal. 1993. "NMDA antagonists partially protect against MPTP induced neurotoxicity in mice." *Neuroreport* 4:387–390.

Brouillet, E., C. Jacquard, N. Bizat, and D. J. Blum. 2005. "3-Nitropropionic acid: A mitochondrial toxin to uncover physiopathological mechanisms underlying striatal degeneration in Huntington's disease." *Journal of Neurochemistry* 95:1521–1540.

Bruce, A. J., W. Boling, M. S. Kindy, J. Peschon, P. J. Kraemer, M. K. Carpenter, et al. 1996. "Altered neuronal and microglial responses to excitotoxic and ischemic brain injury in mice lacking TNF receptors." *Nature Medicine* 2:788–794.

Bruce-Keller, A. J., J. W. Geddes, P. E. Knapp, R. W. McFall, J. N. Keller, and F. W. Holtsberg. 1999. "Anti-death properties of TNF against metabolic poisoning: Mitochondrial stabilization by MnSOD." *Journal of Neuroimmunology* 93:53–71.

Bruce-Keller, A. J., G. Umberger, R. McFall, and M. P. Mattson. 1999. "Food restriction reduces brain damage and improves behavioral outcome following excitotoxic and metabolic insults." *Annals of Neurology* 45:8–15.

Burek, M. J., and R. W. Oppenheim. 1996. "Programmed cell death in the developing nervous system." *Brain Pathology* 6:427–446.

Camandola, S., N. Plick, and M. P. Mattson. 2019. "Impact of coffee and cacao purine metabolites on neuroplasticity and neurodegenerative disease." *Neurochemical Research* 44:214–227.

Castelhano, J., G. Lima, M. Teixeira, C. Soares, M. Pais, and M. Castelo-Branco. 2021. "The effects of tryptamine psychedelics in the brain: A meta-analysis of functional and review of molecular imaging studies." *Frontiers in Pharmacology* 12. doi: 10.3389/fphar.2021.739053.

Castren, E., and M. Kojima. 2017. "Brain-derived neurotrophic factor in mood disorders and antidepressant treatments." *Neurobiology of Disease* 97:119–126.

Catledge, T. 1971. *My Life and Times*. New York: Harper and Row.

Chang, D. T., G. L. Rintoul, S. Pandipati, and I. J. Reynolds. 2006. "Mutant huntingtin aggregates impair mitochondrial movement and trafficking in cortical neurons." *Neurobiology of Disease* 22:388–400.

Chartoff, E. H., and H. S. Connery. 2014. "It's MORe exciting than mu: Crosstalk between mu opioid receptors and glutamatergic transmission in the mesolimbic dopamine system." *Frontiers in Pharmacology* 5. doi: 10.3389/fphar.2014.00116.

Chaves, C., P. C. T. Bittencourt, and A. Pelegrini. 2020. "Ingestion of a THC-rich cannabis oil in people with fibromyalgia: A randomized, double-blind, placebo-controlled clinical trial." *Pain Medicine* 21:2212–2218.

Cheng, A., R. Wan, J. L. Yang, N. Kamimura, T. G. Son, X. Ouyang, et al. 2012. "Involvement of PGC-1alpha in the formation and maintenance of neuronal dendritic spines." *Nature Communications* 3:1250.

Cheng, A., J. Wang, N. Ghena, Q. Zhao, I. Perone, T. M. King, et al. 2020. "SIRT3 haploinsufficiency aggravates loss of GABAergic interneurons and neuronal network hyperexcitability in an Alzheimer's disease model." *Journal of Neuroscience* 40:694–709.

Cheng, A., Y. Yang, Y. Zhou, C. Maharana, D. Lu, W. Peng, et al. 2016. "Mitochondrial SIRT3 mediates adaptive responses of neurons to exercise and metabolic and excitatory challenges." *Cell Metabolism* 23:128–142.

Cheng, B., S. Christakos, and M. P. Mattson. 1994. "Tumor necrosis factors protect neurons against metabolic-excitotoxic insults and promote maintenance of calcium homeostasis." *Neuron* 12:139–153.

Cheng, B., and M. P. Mattson. 1992a. "Glucose deprivation elicits neurofibrillary tangle-like antigenic changes in hippocampal neurons: Prevention by NGF and bFGF." *Experimental Neurology* 117:114–123.

Cheng, B., and M. P. Mattson. 1992b. "IGF-I and IGF-II protect cultured hippocampal and septal neurons against calcium-mediated hypoglycemic damage." *Journal of Neuroscience* 12:1558–1566.

Cheng, B., and M. P. Mattson. 1994. "NT-3 and BDNF protect CNS neurons against metabolic/excitotoxic insults." *Brain Research* 640:56–67.

Cherbuin, N., K. Sargent-Cox, M. Fraser, P. Sachdev, and K. J. Anstey. 2015. "Being overweight is associated with hippocampal atrophy: The PATH through Life Study." *International Journal of Obesity* 39:1509–1514.

Choi, D. W., M. Maulucci-Gedde, and A. R. Kriegstein. 1987. "Glutamate neurotoxicity in cortical cell culture." *Journal of Neuroscience* 7:357–368.

Cicchetti, F., J. Drouin-Ouellet, and R. E. Gross. 2009. "Environmental toxins and Parkinson's disease: What have we learned from pesticide-induced animal models?" *Trends in Pharmacological Sciences* 30:475–483.

Cline, H. T., and M. J. Constantine-Paton. 1990. "NMDA receptor agonist and antagonists alter retinal ganglion cell arbor structure in the developing frog retinotectal projection." *Journal of Neuroscience* 10:1197–1216.

Colcombe, S. J., K. I. Erickson, P. E. Scalf, J. S. Kim, R. Prakash, E. McAuley, et al. 2006. "Aerobic exercise training increases brain volume in aging humans." *Journal of Gerontology A: Biological Sciences and Medical Sciences* 61:1166–1170.

Colizzi, M., P. McGuire, R. G. Pertwee, and S. Bhattacharyya. 2016. "Effect of cannabis on glutamate signalling in the brain: A systematic review of human and animal evidence." *Neuroscience Biobehavioral Reviews* 64:359–381.

Collingridge, G. L., S. J. Kehl, and H. McLennan. 1983. "Excitatory amino acids in synaptic transmission in the Schaffer collateral-commissural pathway of the rat hippocampus." *Journal of Physiology* 334:33–46.

Corder, G., D. C. Castro, M. R. Bruchas, and G. Scherrer. 2018. "Endogenous and exogenous opioids in pain." *Annual Review of Neuroscience* 41:453–473.

Cornell-Bell, A. H., S. M. Finkbeiner, M. S. Cooper, and S. J. Smith. 1990. "Glutamate induces calcium waves in cultured astrocytes: Long-range glial signaling." *Science* 247:470–473.

Cortes-Briones, J., P. D. Skosnik, D. Mathalon, J. Cahill, B. Pittman, A. William, et al. 2015. "Delta9-THC disrupts gamma-band neural oscillations in humans." *Neuropsychopharmacology* 40:2124–2134.

Cotman, C. W., and N. C. Berchtold. 2002. "Exercise: A behavioral intervention to enhance brain health and plasticity." *Trends in Neurosciences* 25:295–301.

Cowan, W. M. 2001. "Viktor Hamburger and Rita Levi-Montalcini: The path to the discovery of nerve growth factor." *Annual Review of Neuroscience* 24:551–600.

Cristino, L., T. Bisogno, and V. Marzo. 2020. "Cannabinoids and the expanded endocannabinoid system in neurological disorders." *Nature Reviews Neurology* 16:9–29.

Crook, Z. R., and D. Housman. 2011. "Huntington's disease: Can mice lead the way to treatment?" *Neuron* 69:423–435.

Cunnane, S. C., E. Trushina, C. Morland, A. Prigione, G. Casadesus, Z. B. Andrews, et al. 2020. "Brain energy rescue: An emerging therapeutic concept for neurodegenerative disorders of ageing." *Nature Reviews Drug Discovery* 19:609–633.

Curtis, D. R., J. W. Phillis, and J. C. Watkins. 1960. "The chemical excitation of spinal neurones by certain acidic amino acids." *Journal of Physiology* 150:656–682.

Dakwar, E., F. Levin, C. L. Hart, C. Basaraba, J. Choi, M. Pavlicova, et al. 2020. "A single ketamine infusion combined with motivational enhancement therapy for alcohol use disorder: A randomized midazolam-controlled pilot trial." *American Journal of Psychiatry* 177:125–133.

Dani, J. W., A. Chernjavsky, and S. J. Smith. 1992. "Neuronal activity triggers calcium waves in hippocampal astrocyte networks." *Neuron* 8:429–440.

Davis, A. K., F. S. Barrett, D. G. May, M. P. Cosimano, N. D. Sepeda, M. W. Johnson, et al. 2020. "Effects of psilocybin-assisted therapy on major depressive disorder: A randomized clinical trial." *JAMA Psychiatry* 78:481–489.

De Cabo, R., and M. P. Mattson. 2019. "Effects of intermittent fasting on health, aging, and disease." *New England Journal of Medicine* 381:2541–2551.

Del Tredici, K., and H. Braak. 2016. "Review: Sporadic Parkinson's disease: Development and distribution of alpha-synuclein pathology." *Neuropathology and Applied Neurobiology* 42:33–50.

Devereaux, A. L., S. L. Mercer, and C. W. Cunningham. 2018. "DARK classics in chemical neuroscience: Morphine." *ACS Chemical Neuroscience* 9:2395–2407.

De Vos, C. M. H., N. L. Mason, and K. P. C. Kuypers. 2021. "Psychedelics and neuroplasticity: A systematic review unraveling the biological underpinnings of psychedelics." *Frontiers in Psychiatry* 12. doi: 10.3389/fpsyt.2021.724606.

Diniz, B. S., A. L. Teixeira, R. Machado-Vieira, L. L. Talib, M. Radanovic, W. F. Gattaz, et al. 2014. "Reduced cerebrospinal fluid levels of brain-derived neurotrophic factor is associated with cognitive impairment in late-life major depression." *Journal of Gerontology B: Psychological Science and Social Science* 69:845–851.

D'Souza, M. S., and A. Markou. 2013. "The 'stop' and 'go' of nicotine dependence: Role of GABA and glutamate." *Cold Spring Harbor Perspectives in Medicine* 3. doi: 10.1101/cshperspect.a012146.

Duan, W., Z. Guo, H. Jiang, B. Ladenheim, X. Xu, J. L. Cadet, et al. 2004. "Paroxetine retards disease onset and progression in huntingtin mutant mice." *Annals of Neurology* 55:590–594.

Duan, W., Z. Guo, H. Jiang, M. Ware, X. J. Li, and M. P. Mattson. 2003. "Dietary restriction normalizes glucose metabolism and BDNF levels, slows disease progression, and increases survival in huntingtin mutant mice." *Proceedings of the National Academy of Science USA* 100:2911–2916.

Duan, W., and M. P. Mattson. 1999. "Dietary restriction and 2-deoxyglucose administration improve behavioral outcome and reduce degeneration of dopaminergic neurons in models of Parkinson's disease." *Journal of Neuroscience Research* 57:195–206.

Dumas, T. C., T. Gillette, D. Ferguson, K. Hamilton, and R. M. Sapolsky. 2010. "Anti-glucocorticoid gene therapy reverses the impairing effects of elevated corticosterone on spatial memory, hippocampal neuronal excitability, and synaptic plasticity." *Journal of Neuroscience* 30:1712–1720.

Dupuy, M., and S. Chanraud. 2016. "Imaging the addicted brain: Alcohol." *International Review of Neurobiology* 129:1–31.

Dwivedi, Y. 2010. "Brain-derived neurotrophic factor and suicide pathogenesis." *Annals of Medicine* 42:87–96.

Elliott, E. M., M. P. Mattson, P. Vanderklish, G. Lynch, I. Chang, and R. M. Sapolsky. 1993. "Corticosterone exacerbates kainate-induced alterations in hippocampal Tau immunoreactivity and spectrin proteolysis in vivo." *Journal of Neurochemistry* 61:57–67.

Emerich, D. F., J. H. Kordower, Y. Chu, C. Thanos, B. Bintz, G. Paolone, et al. 2019. "Widespread striatal delivery of GDNF from encapsulated cells prevents the anatomical and functional consequences of excitotoxicity." *Neural Plasticity*, March 11. doi: 10.1155/2019/6286197.

Enquist, B. J., and K. J. Niklas. 2001. "Invariant scaling relations across tree-dominated communities." *Nature* 410:655–660.

Estes, M. L., and A. K. McAllister. 2016. "Maternal immune activation: Implications for neuropsychiatric disorders." *Science* 353:772–777.

Farmer, J., X. Zhao, H. van Praag, K. Wodtke, F. Gage, and B. R. Christie. 2004. "Effects of voluntary exercise on synaptic plasticity and gene expression in the dentate gyrus of adult male Sprague-Dawley rats in vivo." *Neuroscience* 124:71–79.

Fields, H. L., and E. B. Margolis. 2015. "Understanding opioid reward." *Trends in Neurosciences* 38:217–225.

Fischer, S. 2021. "The hypothalamus in anxiety disorders." *Handbook of Clinical Neurology* 180:149–160.

Fischetti, Mark. 2011. "Computers versus brains." *Scientific American* 305:104.

Forde, B. G. 2014. "Glutamate signalling in roots." *Journal of Experimental Botany* 65:779–787.

Forsse, A., T. H. Nielsen, K. H. Nygaard, C. H. Nordstrom, J. B. Gramsbergen, and F. R. Poulsen. 2019. "Cyclosporin A ameliorates cerebral oxidative metabolism and infarct size in the endothelin-1 rat model of transient cerebral ischaemia." *Science Reports* 9 (March 6): 3702.

Fountain, S. J. 2010. "Neurotransmitter receptor homologues of *Dictyostelium discoideum*." *Molecular Neuroscience* 41:263–266.

Fragoso, Y. D., A. Carra, and M. A. Macias. 2020. "Cannabis and multiple sclerosis." *Expert Reviews in Neurotherapeutics* 20:849–854.

Franklin, T. R., D. Harper, K. Kampman, S. Kildea-McCrea, W. Jens, K. G. Lynch, et al. 2009. "The GABA B agonist baclofen reduces cigarette consumption in a preliminary double-blind placebo-controlled smoking reduction study." *Drug and Alcohol Dependence* 103:30–36.

Fredrikson, M., and V. Faria. 2013. "Neuroimaging in anxiety disorders." *Trends in Pharmacopsychiatry* 29:47–66.

Frye, R. E., M. F. Casanova, S. H. Fatemi, T. D. Folsom, T. J. Reutiman, G. L. Brown, et al. 2016. "Neuropathological mechanisms of seizures in autism spectrum disorder." *Frontiers in Neuroscience* 10. doi: 10.3389/fnins.2016.00192.

Furukawa, K., S. W. Barger, E. M. Blalock, and M. P. Mattson. 1996. "Activation of K+ channels and suppression of neuronal activity by secreted beta-amyloid-precursor protein." *Nature* 379:74–78.

Gallo, G., F. B. Lefcort, and P. C. Letourneau. 1997. "The trkA receptor mediates growth cone turning toward a localized source of nerve growth factor." *Journal of Neuroscience* 17:5445–5454.

Gatt, J. M., K. L. Burton, L. M. Williams, and P. R. Schofield. 2015. "Specific and common genes implicated across major mental disorders: A review of meta-analysis studies." *Journal of Psychiatric Research* 60:1–13.

Gautier, H. O., K. A. Evans, K. Volbracht, R. James, S. Sitnikov, I. Lundgaard, et al. 2015. "Neuronal activity regulates remyelination via glutamate signalling to oligodendrocyte progenitors." *Nature Communications* 6. doi: 10.1038/ncomms9518.

Gillespie, C. 2022. "Lady Gaga developed PTSD after she was 'repeatedly' raped at 19." *Health*, May 17.

Glasper, E. R., M. V. Llorens-Martin, B. Leuner, E. Gould, and J. L. Trejo. 2010. "Blockade of insulin-like growth factor-I has complex effects on structural plasticity in the hippocampus." *Hippocampus* 20:706–712.

Glenny, R. W. 2011. "Emergence of matched airway and vascular trees from fractal rules." *Journal of Applied Physiology* 110:1119–1129.

Goldstein, R. Z., and N. D. Volkow. 2011. "Dysfunction of the prefrontal cortex in addiction: Neuroimaging findings and clinical implications." *Nature Reviews Neuroscience* 12:652–669.

Golgi, C. 1886. *Sulla fina anatomia degli organi centrali del sistema nervoso.* Milan: Hoepli.

Goodrick, C. L., D. K. Ingram, M. A. Reynolds, J. R. Freeman, and N. L. Cider. 1982. "Effects of intermittent feeding upon growth and life span in rats." *Gerontology* 28:233–241.

Gould, E., and P. Tanapat. 1999. "Stress and hippocampal neurogenesis." *Biological Psychiatry* 46:1472–1479.

Griffioen, K. J., S. M. Rothman, B. Ladenheim, R. Wan, N. Vranis, E. Hutchison, et al. 2013. "Dietary energy intake modifies brainstem autonomic dysfunction caused by mutant α-synuclein." *Neurobiology of Aging* 34:928–935.

Griffiths, R. R., M. W. Johnson, M. A. Carducci, A. Umbricht, W. A. Richards, M. P Cosimano, et al. 2016. "Psilocybin produces substantial and sustained decreases in depression and anxiety in patients with life-threatening cancer: A randomized double-blind trial." *Journal of Psychopharmacology* 30:1181–1197.

Grillo, C. A., G. G. Piroli, R. C. Lawrence, S. A. Wrighten, A. J. Green, S. P. Wilson, et al. 2015. "Hippocampal insulin resistance impairs spatial learning and synaptic plasticity." *Diabetes* 64:3927–3936.

Grossman, R. G., M. G. Fehlings, R. F. Frankowski, K. D. Burau, D. S. Chow, C. Tator, et al. 2014. "A prospective, multicenter, phase I matched-comparison group trial of safety, pharmacokinetics, and preliminary efficacy of riluzole in patients with traumatic spinal cord injury." *Journal of Neurotrauma* 31:239–255.

Grynkiewicz, G., M. Poenie, and R. Y. Tsien. 1985. "A new generation of Ca2+ indicators with greatly improved fluorescence properties." *Journal of Biological Chemistry* 260:3440–3450.

Guillemot-Legris, O., and G. G. Muccioli. 2017. "Obesity-induced neuroinflammation: Beyond the hypothalamus." *Trends in Neurosciences* 40:237–253.

Guo, Q., W. Fu, B. L. Sopher, M. W. Miller, C. B. Ware, G. M. Martin, and M. P. Mattson. 1999. "Increased vulnerability of hippocampal neurons to excitotoxic necrosis in presenilin-1 mutant knock-in mice." *Nature Medicine* 5:101–106.

Guthrie, P. B., M. Segal, and S. B. Kater. 1991. "Independent regulation of calcium revealed by imaging dendritic spines." *Nature* 354:76–81.

Halagappa, V. K., Z. Guo, M. Pearson, Y. Matsuoka, R. G. Cutler, F. M. LaFerla, et al. 2007. "Intermittent fasting and caloric restriction ameliorate age-related behavioral deficits in the triple-transgenic mouse model of Alzheimer's disease." *Neurobiology of Disease* 26:212–220.

Hamasaka, Y., D. Rieger, M. L. Parmentier, Y. Grau, C. Helfrich-Forster, and D. R. Nassel. 2007. "Glutamate and its metabotropic receptor in Drosophila clock neuron circuits." *Journal of Comparative Neurology* 505:32–45.

Hao, S., A. Dey, X. Yu, and A. M. Stranahan. 2016. "Dietary obesity reversibly induces synaptic stripping by microglia and impairs hippocampal plasticity." *Brain Behavior and Immunity* 51:230–239.

Harris, G. C., M. Wimmer, R. Byrne, and G. Aston-Jones. 2004. "Glutamate-associated plasticity in the ventral tegmental area is necessary for conditioning environmental stimuli with morphine." *Neuroscience* 129:841–847.

Hartley, T., C. Lever, N. Burgess, and J. O'Keefe. 2013. "Space in the brain: How the hippocampal formation supports spatial cognition." *Philosophical Transactions of the Royal Society London B: Biological Sciences* 369. doi: 10.1098/rstb.2012.0510.

Hebb, D. O. 1949. *The Organization of Behavior*. New York: Wiley and Sons.

Hertz, L. 2013. "The glutamate-glutamine (GABA) cycle: Importance of late postnatal development and potential reciprocal interactions between biosynthesis and degradation." *Frontiers in Endocrinology* 27:59.

Hidese, S., M. Ota, J. Matsuo, I. Ishida, M. Hiraishi, S. Yoshida, et al. 2018. "Association of obesity with cognitive function and brain structure in patients with major depressive disorder." *Journal of Affective Disorders* 225:188–194.

Hofmann, A. 1980. *LSD: My Problem Child*. New York: McGraw-Hill Book Company.

Hofmann, A., A. Frey, H. Ott, T. Petrzilka, and F. Troxler. 1958. "Elucidation of the structure and the synthesis of psilocybin." *Experientia* 14:397–399.

Horner, A. J., J. A. Bisby, E. Zotow, D. Bush, and N. Burgess. 2016. "Grid-like processing of imagined navigation." *Current Biology* 26:842–847.

Hou, Y., Y. Wei, S. Lautrup, B. Yang, Y. Wang, S. Cordonnier, et al. 2021. "NAD+ supplementation reduces neuroinflammation and cell senescence in a transgenic mouse model of Alzheimer's disease via cGAS-STING." *Proceedings of the National Academy of Science USA* 118. doi: 10.1073/pnas.2011226118.

Hüls, S., T. Högen, N. Vassallo, K. M. Danzer, B. Hengerer, A. Giese, et al. 2011. "AMPA-receptor-mediated excitatory synaptic transmission is enhanced by iron-induced alpha-synuclein oligomers." *Journal of Neurochemistry* 117:868–878.

Hunsberger, H. C., D. S. Weitzner, C. C. Rudy, J. E. Hickman, E. M. Libell, R. R. Speer, et al. 2015. "Riluzole rescues glutamate alterations, cognitive deficits, and Tau pathology associated with P301L Tau expression." *Journal of Neurochemistry* 135:381–394.

Huxley, A. 1954. *The Doors of Perception*. New York: Harper-Collins Publishers.

Intlekofer, K. A., and C. W. Cotman. 2013. "Exercise counteracts declining hippocampal function in aging and Alzheimer's disease." *Neurobiology of Disease* 57:47–55.

Ito, M. 2001. "Cerebellar long-term depression: Characterization, signal transduction, and functional roles." *Physiological Reviews* 81:1143–1195.

Jacob, F. *The Possible and the Actual*. New York: Pantheon Books, 1982.

Jayakumar, R. P., M. S. Madhav, F. Savelli, H. T. Blair, N. J. Cowan, and J. J. Knierim. 2019. "Recalibration of path integration in hippocampal place cells." *Nature* 566:533–537.

Jin, H., Y. Zhu, Y. Li, X. Ding, W. Ma, X. Han, et al. 2019. "BDNF-mediated mitophagy alleviates high-glucose-induced brain microvascular endothelial cell injury." *Apoptosis* 24:511–528.

Josselyn, S. A., S. Kohler, and P. W. Frankland. 2017. "Heroes of the engram." *Journal of Neuroscience* 37:4647–4657.

Josselyn, S. A., and S. Tonegawa. 2020. "Memory engrams: Recalling the past and imagining the future." *Science* 367. doi: 10.1126/science.aaw4325.

Kalivas, P. W. 2007. "Cocaine and amphetamine-like psychostimulants: Neurocircuitry and glutamate neuroplasticity." *Dialogues in Clinical Neuroscience* 9:389–397.

Kalivas, P. W., and N. D. Volkow. 2005. "The neural basis of addiction: A pathology of motivation and choice." *American Journal of Psychiatry* 162:1403–1413.

Kang, J., H. G. Lemaire, A. Unterbeck, J. M. Salbaum, C. L. Masters, K. H. Grzeschik, et al. 1987. "The precursor of Alzheimer's disease amyloid A4 protein resembles a cell-surface receptor." *Nature* 325:733–736.

Kano, T., P. J. Brockie, T. Sassa, H. Fujimoto, Y. Kawahara, Y. Iino, et al. 2008. "Memory in *Caenorhabditis elegans* is mediated by NMDA-type ionotropic glutamate receptors." *Current Biology* 18:1010–1015.

Kantrowitz, J. T., S. W. Woods, E. Petkova, B. Cornblatt, C. M. Corcoran, H. Chen, et al. 2015. "D-serine for the treatment of negative symptoms in individuals at clinical high risk of schizophrenia: A pilot, double-blind, placebo-controlled, randomised parallel group mechanistic proof-of-concept trial." *Lancet Psychiatry* 2:403–412.

Kapogiannis, D., A. Boxer, J. B. Schwartz, E. L. Abner, A. Biragyn, U. Masharani, et al. 2015. "Dysfunctionally phosphorylated type 1 insulin receptor substrate in neural-derived blood exosomes of preclinical Alzheimer's disease." *FASEB Journal* 29:589–596.

Kashiwaya, Y., C. Bergman, J. H. Lee, R. Wan, M. T. King, M. R. Mughal, et al. 2013. "A ketone ester diet exhibits anxiolytic and cognition-sparing properties, and lessens amyloid and Tau pathologies in a mouse model of Alzheimer's disease." *Neurobiology of Aging* 34:1530–1539.

Katz, B., and R. Miledi. 1965. "The effect of calcium on acetylcholine release from motor nerve terminals." *Proceedings of the Royal Society of London B: Biological Sciences* 161:496–503.

Katz, L. C., and C. J. Shatz. 1996. "Synaptic activity and the construction of cortical circuits." *Science* 274:1133–1138.

Keinanen, K., W. Wisden, B. Sommer, P. Werner, A. Herb, T. A. Verdoorn, et al. 1990. "A family of AMPA-selective glutamate receptors." *Science* 249:556–560.

Kempermann, G. 2019. "Environmental enrichment, new neurons and the neurobiology of individuality." *Nature Reviews Neuroscience* 20:235–245.

Kenney, K., F. Amyot, C. Moore, M. Haber, L. C. Turtzo, C. Shenouda, et al. 2018. "Phosphodiesterase-5 inhibition potentiates cerebrovascular reactivity in chronic traumatic brain injury." *Annals of Clinical and Translational Neurology* 5:418–428.

Kim, C. K., A. Adhikar, and K. Deisseroth. 2017. "Integration of optogenetics with complementary methodologies in systems neuroscience." *Nature Reviews Neuroscience* 18:222–235.

Kim, G., O. Gautier, E. Tassoni-Tsuchida, X. R. Ma, and A. D. Gitler. 2020. "ALS genetics: Gains, losses, and implications for future therapies." *Neuron* 108:822–842.

Kim, S., S. H. Kwon, T. I. Kam, N. Panicker, S. S. Karuppagounder, S. Lee, et al. 2019. "Transneuronal propagation of pathologic alpha-synuclein from the gut to the brain models Parkinson's disease." *Neuron* 103:627–641.

Kirwan, P., B. Turner-Bridger, M. Peter, A. Momoh, D. Arambepola, and H. P. Robinson. 2015. "Development and function of human cerebral cortex neural networks from pluripotent stem cells in vitro." *Development* 142:3178–3187.

Kishimoto, Y., J. Johnson, W. Fang, J. Halpern, K. Marosi, J. G. Geisler, et al. 2020. "A mitochondrial uncoupler prodrug protects dopaminergic neurons and improves functional outcome in a mouse model of Parkinson's disease." *Neurobiology of Aging* 85:123–130.

Kishimoto, Y., W. Zhu, W. Hsoda, J. M. Sen, and M. P. Mattson. 2019. "Chronic mild gut inflammation accelerates brain neuropathology and motor dysfunction in alpha-synuclein mutant mice." *Neuromolecular Medicine* 21:239–249.

Koul, O. 2005. *Insect Antifeedants*. New York: Taylor and Francis.

Krafft, C. E., N. F. Schwartz, L. Chi, A. L. Weinberger, D. J. Schaeffer, J. E. Pierce, et al. 2014. "An 8-month randomized controlled exercise trial alters brain activation during cognitive tasks in overweight children." *Obesity* 22:232–242.

Krogsgaard-Larsen, P., T. Honore, J. J. Hansen, D. R. Curtis, and D. Lodge. 1980. "New class of glutamate agonist structurally related to ibotenic acid." *Nature* 284:64–66.

Kuroki, S., Y. Takamasa, T. Hidekazu, M. Iwama, R. Ando, T. Michikawa, et al. 2018. "Excitatory neuronal hubs configure multisensory integration of slow waves in association cortex." *Cell Reports* 22:2873–2885.

Kushner, L., M. V. Bennett, and R. S. Zukin. 1993. "Molecular biology of PCP and NMDA receptors." *NIDA Research Monographs* 133:159–183.

Lacerda-Pinheiro, S. F., R. F. Pinheiro, M. A. Pereira de Lima, C. G. Lima da Silva, S. Vieira dos Santos, A. G. Teixeira, et al. 2014. "Are there depression and anxiety genetic markers and mutations? A systematic review." *Journal of Affective Disorders* 168:387–398.

Lambert, K., M. Hyer, M. Bardi, A. Rzucidlo, S. Scott, B. Terhune-Cotter, et al. 2016. "Natural-enriched environments lead to enhanced environmental engagement and altered neurobiological resilience." *Neuroscience* 330:386–394.

Langston, J. W. 2017. "The MPTP story." *Journal of Parkinson's Disease* 7:S11–S19.

Lanzillotta, C., F. Di Domenico, M. Perluigi, and D. A. Butterfield. 2019. "Targeting mitochondria in Alzheimer's disease: Rationale and perspectives." *CNS Drugs* 33:957–969.

Le Berre, A. P., G. Rauchs, R. La Joie, F. Mezenge, C. Boudehent, F. Vabret, et al. 2014. "Impaired decision-making and brain shrinkage in alcoholism." *European Psychiatry* 29:125–133.

Leblanc, R. 2021. "The birth of experimental neurosurgery: Wilder Penfield at Montreal's Royal Victoria Hospital, 1928–1934." *Journal of Neurosurgery* 136:553–560.

Lee, J., J. P. Herman, and M. P. Mattson. 2000. "Dietary restriction selectively decreases glucocorticoid receptor expression in the hippocampus and cerebral cortex of rats." *Experimental Neurology* 166:435–441.

Leger, M., A. Quiedeville, E. Paizanis, S. Natkunarajah, T. Freret, M. Boulourd, et al. 2012. "Environmental enrichment enhances episodic-like memory in association with a modified neuronal activation profile in adult mice." *PLoS One* 7. doi: 10.1371/journal.pone.0048043.

Leitao, N., P. Dangeville, R. Carter, and M. Charpentier. 2019. "Nuclear calcium signatures are associated with root development." *Nature Communications* 10. doi: 10.1038/s41467-019-12845-8.

Li, A. K., M. J. Koroly, M. E. Schattenkerk, R. A. Malt, and M. Young. 1980. "Nerve growth factor: Acceleration of the rate of wound healing in mice." *Proceedings of the National Academy of Sciences USA* 77:4379–4381.

Li, L., W. W. Men, Y. K. Chang, M. X. Fan, L. Ji, and G. X. Wei. 2014. "Acute aerobic exercise increases cortical activity during working memory: A functional MRI study in female college students." *PLoS One* 9. doi: 10.1371/journal.pone.0099222.

Li, W., X. Xu, and L. Pozzo-Miller. 2016. "Excitatory synapses are stronger in the hippocampus of Rett syndrome mice due to altered synaptic trafficking of AMPA-type glutamate receptors." *Proceedings of the National Academy of Science USA* 113:E1575–E1584.

Li, Y., T. Perry, M. S. Kindy, B. K. Harvey, D. Tweedie, H. W. Holloway, et al. 2009. "GLP-1 receptor stimulation preserves primary cortical and dopaminergic neurons in cellular and rodent models of stroke and Parkinsonism." *Proceedings of the National Academy of Science USA* 106:1285–1290.

Li, Z., K. Okamoto, Y. Hayashi, and M. Sheng. 2014. "The importance of mitochondria in the morphogenesis and plasticity of spines and synapses." *Cell* 119:873–887.

Lin, L., R. Osan, and J. Z. Tsien. 2006. "Organizing principles of real-time memory encoding: Neural clique assemblies and universal neural codes." *Trends in Neurosciences* 29:48–57.

Liu, D., H. Lu, E. Stein, Z. Zhou, Y. Yang, and M. P. Mattson. 2018. "Brain regional synchronous activity predicts tauopathy in 3×TgAD mice." *Neurobiology of Disease* 70:160–169.

Liu, D., M. Pitta, J. H. Lee, B. Ray, D. K. Lahiri, K. Furukawa, et al. 2010. "The KATP channel activator diazoxide ameliorates amyloid-beta and Tau pathologies and improves memory in the 3xTgAD mouse model of Alzheimer's disease." *Journal of Alzheimer's Disease* 22:443–457.

Liu, H., C. Zhang, J. Xu, J. Jin, L. Cheng, X. Miao, et al. 2021. "Huntingtin silencing delays onset and slows progression of Huntington's disease: A biomarker study." *Brain* 144:3101–3113.

Liu, X., S. Ramirez, P. T. Pang, C. B. Puryear, A. Govindarajan, K. Deisseroth, et al. 2012. "Optogenetic stimulation of a hippocampal engram activates fear memory recall." *Nature* 484:381–385.

Liu, Y., A. Cheng, Y. J. Li, Y. Yang, Y. Kishimoto, S. Zhang, et al. 2019. "SIRT3 mediates hippocampal synaptic adaptations to intermittent fasting and ameliorates deficits in APP mutant mice." *Nature Communications* 10:1886.

Lodge, D., J. C. Watkins, Z. A. Bortolotto, D. E. Jane, and A. Volianskis. 2019. "The 1980s: D-AP5, LTP and a decade of NMDA receptor discoveries." *Neurochemical Research* 44:516–530.

Lomo, T. 2003. "The discovery of long-term potentiation." *Philosophical Transactions of the Royal Society London B: Biological Sciences* 358:617–620.

López-Cruz, A., A. Sordillo, N. Pokala, Q. Liu, P. T. McGrath, and C. L. Bargmann. 2019. "Parallel multimodal circuits control an innate foraging behavior." *Neuron* 102:407–419.

Lorenzetti, V., E. Hoch, and W. Hall. 2020. "Adolescent cannabis use, cognition, brain health and educational outcomes: A review of the evidence." *European Neuropsychopharmacology* 36:169–180.

Ludolph, A. C., F. He, P. S. Spencer, J. Hammerstad, and M. Sabri. 1991. "3-Nitropropionic acid-exogenous animal neurotoxin and possible human striatal toxin." *Canadian Journal of Neurological Science* 18:492–498.

Ly, C., A. C. Greb, L. P. Cameron, J. M. Wong, E. V. Barragan, P. C. Wilson, et al. 2018. "Psychedelics promote structural and functional neural plasticity." *Cell Reports* 23:3170–3182.

Maguire, E. A., D. G. Gadian, I. S. Johnsrude, C. D. Good, J. Ashburner, R. Frackowiak, et al. 2000. "Navigation-related structural change in the hippocampi of taxi drivers." *Proceedings of the National Academy of Sciences USA* 97:4398–4403.

Maltbie, E. A., G. S. Kaundinya, and L. L. Howell. 2017. "Ketamine and pharmacological imaging: Use of functional magnetic resonance imaging to evaluate mechanisms of action." *Behavioral Pharmacology* 28:610–622.

Marco, S., A. Giralt, M. M. Petrovic, M. A. Pouladi, R. Martinez-Turrillas, J. Hernandez, et al. 2013. "Suppressing aberrant GluN3A expression rescues synaptic and behavioral impairments in Huntington's disease models." *Nature Medicine* 19:1030–1038.

Marini, A. M., and S. M. Paul. 1992. "N-methyl-D-aspartate receptor-mediated neuroprotection in cerebellar granule cells requires new RNA and protein synthesis." *Proceedings of the National Academy of Science USA* 89:6555–6559.

Mark, R. J., M. A. Lovell, W. R. Markesbery, K. Uchida, and M. P. Mattson. 1997. "A role for 4-hydroxynonenal, an aldehydic product of lipid peroxidation, in disruption of ion homeostasis and neuronal death induced by amyloid beta-peptide." *Journal of Neurochemistry* 68:255–264.

Mark, R. J., Z. Pang, J. W. Geddes, K. Uchida, and M. P. Mattson. 1997. "Amyloid beta peptide impairs glucose transport in hippocampal and cortical neurons: Involvement of membrane lipid peroxidation." *Journal of Neuroscience* 17:1046–1054.

Marosi, K., and M. P. Mattson. 2014. "BDNF mediates adaptive brain and body responses to energetic challenges." *Trends in Endocrinology and Metabolism* 25:89–98.

Martin, A., J. N. Booth, Y. Laird, J. Sproule, J. J. Reilly, and D. H. Saunders. 2018. "Physical activity, diet and other behavioural interventions for improving cognition and school achievement in children and adolescents with obesity or overweight." *Cochrane Database Systems Review* 3. doi: 10.1002/14651858.CD009728.pub3.

Martin, B., E. Golden, O. D. Carlson, P. Pistell, J. Zhou, W. Kim, et al. 2009. "Exendin-4 improves glycemic control, ameliorates brain and pancreatic pathologies, and extends survival in a mouse model of Huntington's disease." *Diabetes* 58:318–328.

Martin, D. A., and C. D. Nichols. 2016. "Psychedelics recruit multiple cellular types and produce complex transcriptional responses within the brain." *EBioMedicine* 11:262–277.

Mason, N. L., K. P. C. Kuypers, F. Muller, J. Reckweg, D. H. Y. Tse, S. W. Toennes, et al. 2020. "Me, myself, bye: Regional alterations in glutamate and the experience of ego dissolution with psilocybin." *Neuropsychopharmacology* 45:2003–2011.

Mattson, M. P. 1990. "Antigenic changes similar to those seen in neurofibrillary tangles are elicited by glutamate and Ca2+ influx in cultured hippocampal neurons." *Neuron* 4:105–117.

Mattson, M. P. 2004. "Pathways towards and away from Alzheimer's disease." *Nature* 430:631–639.

Mattson, M. P. 2012. "Energy intake and exercise as determinants of brain health and vulnerability to injury and disease." *Cell Metabolism* 16:706–722.

Mattson, M. P. 2014. "Superior pattern processing is the essence of the evolved human brain." *Frontiers in Neuroscience* 8. doi: 10.3389/fnins.2014.00265.

Mattson, M. P. 2015. "WHAT DOESN'T KILL YOU." *Scientific American* 313:40–45.

Mattson, M. P. 2019. "An evolutionary perspective on why food overconsumption impairs cognition." *Trends in Cognitive Science* 23:200–212.

Mattson, M. P. 2020. "Involvement of GABAergic interneuron dysfunction and neuronal network hyperexcitability in Alzheimer's disease: Amelioration by metabolic switching." *International Review of Neurobiology* 154:191–205.

Mattson, M. P. 2021. "Applying available knowledge and resources to alleviate familial and sporadic neurodegenerative disorders." *Progress in Molecular Biology and Translational Science* 177:91–107.

Mattson, M. P. 2022. *The Intermittent Fasting Revolution: The Science of Optimizing Health and Enhancing Performance*. Cambridge, MA: MIT Press.

Mattson, M. P., and T. V. Arumugam. 2018. "Hallmarks of brain aging: Adaptive and pathological modification by metabolic states." *Cell Metabolism* 27:1176–1199.

Mattson M. P., B. Cheng, A. R. Culwell, F. S. Esch, I. Lieberburg, and R. E. Rydel. 1993. "Evidence for excitoprotective and intraneuronal calcium-regulating roles for secreted forms of the β-amyloid precursor protein." *Neuron* 10:243–254.

Mattson, M. P., B. Cheng, D. Davis, K. Bryant, I. Lieberburg, and R. E. Rydel. 1992. "Beta-amyloid peptides destabilize calcium homeostasis and render human cortical neurons vulnerable to excitotoxicity." *Journal of Neuroscience* 12:376–389.

Mattson, M. P., P. Dou, and S. B. Kater. 1988. "Outgrowth-regulating actions of glutamate in isolated hippocampal pyramidal neurons." *Journal of Neuroscience* 8:2087–2100.

Mattson, M. P., and S. B. Kater. 1989. "Development and selective neurodegeneration in cell cultures from different hippocampal regions." *Brain Research* 490:110–125.

Mattson, M. P., J. N. Keller, and J. G. Begley. 1998. "Evidence for synaptic apoptosis." *Experimental Neurology* 153:35–48.

Mattson, M. P., R. E. Lee, M. E. Adams, P. B. Guthrie, and S. B. Kater. 1988. "Interactions between entorhinal axons and target hippocampal neurons: A role for glutamate in the development of hippocampal circuitry." *Neuron* 1:865–876.

Mattson, M. P., M. A. Lovell, K. Furukawa, and W. R. Markesbery. 1995. "Neurotrophic factors attenuate glutamate-induced accumulation of peroxides, elevation of intracellular Ca2+ concentration, and neurotoxicity and increase antioxidant enzyme activities in hippocampal neurons." *Journal of Neurochemistry* 65:1740–1751.

Mattson, M. P., K. Moehl, N. Ghena, M. Schmaedick, and A. Cheng. 2018. "Intermittent metabolic switching, neuroplasticity and brain health." *Nature Reviews Neuroscience* 19:63–80.

Mattson, M. P., M. Murrain, P. B. Guthrie, and S. B. Kater. 1989. "Fibroblast growth factor and glutamate: Opposing roles in the generation and degeneration of hippocampal neuroarchitecture." *Journal of Neuroscience* 9:3728–3740.

Mattson, M. P., B. Rychlik, C. Chu, and S. Christakos. 1991. "Evidence for calcium-reducing and excitoprotective roles for the calcium-binding protein calbindin-D28k in cultured hippocampal neurons." *Neuron* 6:41–51.

McEwen, B. S., N. P. Bowles, J. D. Gray, M. N. Hill, R. G. Hunter, I. N. Karatsoreos, and C. Nasca. 2015. "Mechanisms of stress in the brain." *Nature Neuroscience* 18:1353–1363.

McEwen, B. S., L. Eiland, R. G. Hunter, and M. M. Miller. 2012. "Stress and anxiety: Structural plasticity and epigenetic regulation as a consequence of stress." *Neuropharmacology* 62: 3–12.

McKee, A. C. 2020. "The neuropathology of chronic traumatic encephalopathy: The status of the literature." *Seminars in Neurology* 40:359–369.

McLennan, H. 1974. "Actions of excitatory amino acids and their antagonism." *Neuropharmacology* 13:449–454.

McShane, R., M. J. Westby, E. Roberts, N. Minakaran, L. Schneider, L. E. Farrimond, et al. 2019. "Memantine for dementia." *Cochrane Database Systematic Review* 3. doi: 10.1002/14651858 .CD003154.pub6.

Merritt, K., P. McGuire, and A. Egerton. 2013. "Relationship between glutamate dysfunction and symptoms and cognitive function in psychosis." *Frontiers in Psychiatry* 4. doi: 10.3389 /fpsyt.2013.00151.

Miledi, R. 1967. "Spontaneous synaptic potentials and quantal release of transmitter in the stellate ganglion of the squid." *Journal of Physiology* 192:379–406.

Miller, R. G., J. P. Bouchard, P. Duquette, A. Eisen, D. Gelinas, Y. Harati, et al. 1996. "Clinical trials of riluzole in patients with ALS. ALS/Riluzole Study Group-II." *Neurology* 47 (supplement 2): S86–90.

Mindt, S., M. Neumaier, R. Hellweg, A. Sartorius, and L. J. Kranaster. 2020. "Brain-derived neurotrophic factor in the cerebrospinal fluid increases during electroconvulsive therapy in patients with depression: A preliminary report." *Journal of Electroconvulsive Therapy* 36:193–197.

Monteiro, P., and G. Feng. 2017. "SHANK proteins: Roles at the synapse and in autism spectrum disorder." *Nature Reviews of Neuroscience* 18:147–157.

Moriyoshi, K., M. Masu, T. Ishii, R. Shigemoto, N. Mizuno, and S. Nakanishi. 1991. "Molecular cloning and characterization of the rat NMDA receptor." *Nature* 354:31–37.

Morland, C., K. A. Andersson, O. P. Haugen, A. Hadzic, L. Kleppa, A. Gille, et al. 2017. "Exercise induces cerebral VEGF and angiogenesis via the lactate receptor HCAR1." *Nature Communications* 8:1–9.

Moser, M. B., D. C. Rowland, and E. I. Moser. 2015. "Place cells, grid cells, and memory." *Cold Spring Harbor Perspectives on Biology* 7. doi: 10.1101/cshperspect.a021808.

Mostajeran, F., J. Krzikawski, F. Steinicke, and S. Kuhn. 2021. "Effects of exposure to immersive videos and photo slideshows of forest and urban environments." *Science Reports* 11. doi: 10.1038/ s41598-021-83277-y.

Mousavi, S. A. R., A. Chauvin, F. Pascaud, S. Kellenberger, and E. E. Farmer. 2013. "Glutamate receptor-like genes mediate leaf-to-leaf wound signaling." *Nature* 500:422–426.

Mughal, M. R., A. Baharani, S. Chigurupati, T. G. Son, E. Chen, P. Yang, et al. 2011. "Electro-convulsive shock ameliorates disease processes and extends survival in huntingtin mutant mice." *Human Molecular Genetics* 20:659–669.

Murray, M., D. Kim, Y. Liu, C. Tobias, A. Tessler, and I. Fischer. 2002. "Transplantation of genetically modified cells contributes to repair and recovery from spinal injury." *Brain Research Reviews* 40:292–300.

Musaeus, C. S., M. M. Shafi, E. Santarnecchi, S. T. Herman, and D. Z. Press. 2017. "Levetiracetam alters oscillatory connectivity in Alzheimer's disease." *Alzheimer's Disease* 58:1065–1076.

Mustafa, A. K., M. M. Gadalla, and S. H. Snyder. 2009. "Signaling by gasotransmitters." *Science Signaling* 2. doi: 10.1126/scisignal.268re2.

Nair, A., and V. A. Vaidya. 2006. "Cyclic AMP response element binding protein and brain-derived neurotrophic factor: Molecules that modulate our mood?" *Journal of Bioscience* 31:423–434.

Naka, D., and K. R. Mills. 2000. "Further evidence for corticomotor hyperexcitability in amyo-trophic lateral sclerosis." *Muscle and Nerve* 23:1044–1050.

Nees, F., and S. T. Pohlack. 2014. "Functional MRI studies of the hippocampus." *Frontiers in Neurology and Neuroscience* 34:85–94.

Neher, E., and B. Sakmann. 1976. "Single-channel currents recorded from membrane of dener-vated frog muscle fibres." *Nature* 260:799–802.

Neth, B. J., A. Mintz, C. Whitlow, Y. Jung, S. K. Solingapuram, T. C. Register, et al. 2020. "Mod-ified ketogenic diet is associated with improved cerebrospinal fluid biomarker profile, cerebral perfusion, and cerebral ketone body uptake in older adults at risk for Alzheimer's disease: A pilot study." *Neurobiology of Aging* 86:54–63.

Nichols, D. E. 2016. "Psychedelics." *Pharmacological Reviews* 68:264–355.

Nighoghossian, N., Y. Berthezene, L. Mechtouff, L. Derex, T. H. Cho, T. Ritzenthaler, et al. 2015. "Cyclosporine in acute ischemic stroke." *Neurology* 84:2216–2223.

Norwitz, N. G., D. J. Dearlove, M. Lu, K. Clarke, H. Dawes, and M. T. Hu. 2020. "A ketone ester drink enhances endurance exercise performance in Parkinson's disease." *Frontiers in Neuroscience* 14. doi: 10.3389/fnins.2020.584130.

O'Connell, L. A., and H. A. Hofmann. 2011. "The vertebrate mesolimbic reward system and social behavior network: A comparative synthesis." *Journal of Comparative Neurology* 519:3599–3639.

Oddo, S., A. Caccamo, J. D. Shepherd, M. P. Murphy, T. E. Golde, R. Kayed, et al. 2003. "Tri-ple-transgenic model of Alzheimer's disease with plaques and tangles: Intracellular Abeta and synaptic dysfunction." *Neuron* 39:409–421.

Ohline, S. M., and W. C. Abraham. 2019. "Environmental enrichment effects on synaptic and cellular physiology of hippocampal neurons." *Neuropharmacology* 145:3–12.

Olney, J. W. 1989. "Glutamate, a neurotoxic transmitter." *Journal of Childhood Neurology* 4:218–226.

Ortiz-Ramirez, C., E. Michard, A. A. Simon, D. S. C. Damineli, M. Hernandez-Coronado, J. D. Becker, and J. A. Feijo. 2017. "Glutamate receptor-like channels are essential for chemotaxis and reproduction in mosses." *Nature* 549:91–95.

Oskarsson, B., E. A. Mauricio, J. S. Shah, Z. Li, and M. A. Rogawski. 2021. "Cortical excitability threshold can be increased by the AMPA blocker Perampanel in amyotrophic lateral sclerosis." *Muscle and Nerve* 64:215–219.

Paasonen, J., R. A. Salo, J. Ihalainen, J. V. Leikas, K. Savolainen, M. Lehtonen, et al. 2017. "Dose–response effect of acute phencyclidine on functional connectivity and dopamine levels, and their association with schizophrenia-like symptom classes in rat." *Neuropharmacology* 119:15–25.

Patten, A. R., S. Y. Yau, C. J. Fontaine, A. Meconi, R. C. Wortman, and B. R. Christie. 2015. "The benefits of exercise on structural and functional plasticity in the rodent hippocampus of different disease models." *Brain Plasticity* 1:97–127.

Paul, B. D., and S. H. Snyder. 2018. "Gasotransmitter hydrogen sulfide signaling in neuronal health and disease." *Biochemical Pharmacology* 149:101–109.

Pearson, J. M., K. K. Watson, and M. L. Platt. 2014. "Decision making: The neuroethological turn." *Neuron* 82:950–965.

Perry, T., N. J. Haughey, M. P. Mattson, J. M. Egan, and N. H. Greig. 2002. "Protection and reversal of excitotoxic neuronal damage by glucagon-like peptide-1 and exendin-4." *Journal of Pharmacology and Experimental Therapeutics* 302:881–888.

Petralia, R. S., Y. X. Wang, M. P. Mattson, and P. J. Yao. 2016. "The diversity of spine synapses in animals." *Neuromolecular Medicine* 18:497–539.

Pistillo, F., F. Clementi, M. Zoli, and C. Gotti. 2015. "Nicotinic, glutamatergic and dopaminergic synaptic transmission and plasticity in the mesocorticolimbic system: Focus on nicotine effects." *Progress in Neurobiology* 124:1–27.

Plowey, E. D., J. W. Johnson, E. Steer, W. Zhu, D. A. Eisenberg, N. M. Valentino, et al. 2014. "Mutant LRRK2 enhances glutamatergic synapse activity and evokes excitotoxic dendrite degeneration." *Biochimica et Biophysica Acta* 1842:1596–1603.

Polymeropoulos, M. H., C. Lavedan, E. Leroy, S. E. Ide, A. Dehejia, A. Dutra, et al. 1997. "Mutation in the alpha-synuclein gene identified in families with Parkinson's disease." *Science* 276:2045–2047.

Popoli, M., Z. Yan, B. S. McEwen, and G. Sanacora. 2011. "The stressed synapse: The impact of stress and glucocorticoids on glutamate transmission." *Nature Reviews Neuroscience* 13:22–37.

Price, M. B., J. Jelesko, and S. Okumoto. 2012. "Glutamate receptor homologs in plants: Functions and evolutionary origins." *Frontiers in Plant Sciences* 3. doi: 10.3389/fpls.2012.00235.

Purpura, D. P., M. Girado, T. G. Smith, D. A. Callan, and H. Grundfest. 1958. "Structure–activity determinants of pharmacological effects of amino acids and related compounds on central synapses." *Journal of Neurochemistry* 3:238–268.

Puzzo, D., A. Staniszewski, S. X. Deng, L. Privitera, E. Leznik, S. Liu, et al. 2009. "Phosphodies-terase 5 inhibition improves synaptic function, memory, and amyloid-beta load in an Alzheimer's disease mouse model." *Journal of Neuroscience* 29:8075–8086.

Qin, Z., D. Hu, S. Han, S. H. Reaney, D. A. Di Monte, and A. L. Fink. 2007. "Effect of 4-hydroxy-2-nonenal modification on alpha-synuclein aggregation." *Journal of Biological Chemistry* 282:5862–5870.

Qiu, X. M., Y. Y. Sun, X. Y. Ye, and Z. G. Li. 2020. "Signaling role of glutamate in plants." *Frontiers in Plant Science* 10. doi: 10.3389/fpls.2019.01743.

Rabchevsky, A. G., I. Fugaccia, A. F. Turner, D. A. Blades, M. P. Mattson, and S. W. Scheff. 2000. "Basic fibroblast growth factor (bFGF) enhances functional recovery following severe spinal cord injury to the rat." *Experimental Neurology* 164:280–291.

Raefsky, S. M., and M. P. Mattson. 2017. "Adaptive responses of neuronal mitochondria to bio-energetic challenges: Roles in neuroplasticity and disease resistance." *Free Radical Biology and Medicine* 102:203–216.

Ramirez, S., X. Liu, P. A. Lin, J. Suh, M. Pigatelli, R. L. Redondo, et al. 2013. "Creating a false memory in the hippocampus." *Science* 341:387–391.

Renard, J., H. J. Szkudlarek, C. P. Kramar, C. E. L. Jobson, K. Moura, W. J. Rushlow, et al. 2017. "Adolescent THC exposure causes enduring prefrontal cortical disruption of GABAergic inhibition and dysregulation of sub-cortical dopamine function." *Science Reports* 7. doi: 10.1038/s41598-017-11645-8.

Renshaw, D. 2015. "Lady Gaga: 'I've suffered through depression and anxiety my entire life.'" *New Medical Express*, October 15.

Rice, H. C., D. Malmazet, A. Schreurs, S. Frere, I. Van Molle, A. N. Volkov, et al. 2019. "Secreted amyloid-β precursor protein functions as a GABABR1a ligand to modulate synaptic transmission." *Science* 363. doi: 10.1126/science.aao4827.

Richter, M. C., S. Ludewig, A. Winschel, T. Abel, C. Bold, L. R. Salzburger, et al. 2018. "Distinct *in vivo* roles of secreted APP ectodomain variants APPsα and APPsβ in regulation of spine density, synaptic plasticity, and cognition." *EMBO Journal* 37. doi: 10.15252/embj.201798335.

Ring, D., Y. Wolman, N. Friedmann, and S. L. Miller. 1972. "Prebiotic synthesis of hydrophobic and protein amino acids." *Proceedings of the National Academy of Sciences USA* 69:765–768.

Ring, S., S. W. Weyer, S. B. Kililan, E. Waldron, C. U. Pietrzik, M. A. Filippov, et al. 2007. "The secreted β-amyloid precursor protein ectodomain APPs α is sufficient to rescue the anatomical, behavioral and electrophysiological abnormalities of APP-deficient mice." *Journal of Neuroscience* 27:7817–7826.

Rivell, A., and M. P. Mattson. 2019. "Intergenerational metabolic syndrome and neuronal network hyperexcitability in autism." *Trends in Neurosciences* 42:709–726.

Robbins, J. 1958. "The effects of amino acids on the crustacean neuro-muscular system." *Anatomical Record* 132:492–493.

Roberts, J. 2017. "High times." *Distillations* 2:36–39.

Robinson, T. E., and B. Kolb. 1999. "Alterations in the morphology of dendrites and dendritic spines in the nucleus accumbens and prefrontal cortex following repeated treatment with amphetamine or cocaine." *European Journal of Neuroscience* 11:1598–1604.

Rothstein, J. D. 2009. "Current hypotheses for the underlying biology of amyotrophic lateral sclerosis." *Annals of Neurology* 65 (supplement 1): S3–9.

Rothstein, J. D., M. Dykes-Hoberg, C. A. Pardo, L. A. Bristol, L. Jin, R. W. Kuncl, et al. 1996. "Knockout of glutamate transporters reveals a major role for astroglial transport in excitotoxicity and clearance of glutamate." *Neuron* 16:675–686.

Rothstein, J. D., M. Van Kammen, A. I. Levey, L. J. Martin, and R. W. Kuncl. 1995. "Selective loss of glial glutamate transporter GLT-1 in amyotrophic lateral sclerosis." *Annals of Neurology* 38:73–84.

Rubenstein, J. L., and M. M. Merzenich. 2003. "Model of autism: Increased ratio of excitation/inhibition in key neural systems." *Genes Brain and Behavior* 2:255–267.

Salvador, A. F., K. A. de Lima, and J. Kipnis. 2021. "Neuromodulation by the immune system: A focus on cytokines." *Nature Reviews Immunology* 21:526–541.

Sampson, T. R., J. W. Debelius, T. Thron, S. Janssen, G. G. Shastri, Z. E. Iihan, et al. 2016. "Gut microbiota regulate motor deficits and neuroinflammation in a model of Parkinson's disease." *Cell* 167:1469–1480.

Sandhu, K. V., D. Lang, B. Muller, S. Nullmeier, Y. Yanagawa, H. Schwegler, et al. 2014. "Glutamic acid decarboxylase 67 haplodeficiency impairs social behavior in mice." *Genes, Brain and Behavior* 13:439–450.

Scarmeas, N., and Y. Stern. 2004. "Cognitive reserve: Implications for diagnosis and prevention of Alzheimer's disease." *Current Neurology and Neuroscience Reports* 4:374–380.

Schmidt, H. D., and R. S. Duman. 2007. "The role of neurotrophic factors in adult hippocampal neurogenesis, antidepressant treatments and animal models of depressive-like behavior." *Behavioral Pharmacology* 18:391–418.

Schoenfeld, T. J., P. Rada, P. R. Pieruzzini, B. Hsueh, and E. J. Gould. 2013. "Physical exercise prevents stress-induced activation of granule neurons and enhances local inhibitory mechanisms in the dentate gyrus." *Journal of Neuroscience* 33:7770–7777.

Schuman, E. M., and D. V. Madison. 1991. "A requirement for the intercellular messenger nitric oxide in long-term potentiation." *Science* 254:1503–1506.

Schwartz, M., and C. Raposo. 2014. "Protective autoimmunity: A unifying model for the immune network involved in CNS repair." *Neuroscientist* 20:343–358.

Semon, R. 1921. *The Mneme*. London: Allen, Unwin.

Semon, R. W. 1923. *Mnemic Psychology*. London: Allen, Unwin.

Serrano, F., and E. Klann. 2004. "Reactive oxygen species and synaptic plasticity in the aging hippocampus." *Ageing Research Reviews* 3:431–443.

Shao, L. X., C. Liao, I. Gregg, A. P. A. Davoudian, N. K. Savalia, K. Delagarza, et al. 2021. "Psilocybin induces rapid and persistent growth of dendritic spines in frontal cortex in vivo." *Neuron* 109:2535–2544.

Sheline, Y. I., B. M. Disabato, J. Hranilovich, C. Morris, G. D'Angelo, C. Pieper, et al. 2012. "Treatment course with antidepressant therapy in late-life depression." *American Journal of Psychiatry* 169:1185–1193.

Singleton, A. B., M. Farrer, J. Johnson, A. Singleton, S. Hague, J. Kachergus, et al. 2003. "Alpha-synuclein locus triplication causes Parkinson's disease." *Science* 302:841.

Sinz, F. H., X. Pitkow, J. Reimer, M. Bethge, and A. S. Tolias. 2019. "Engineering a less artificial intelligence." *Neuron* 103:967–979.

Slevin, J. T., D. M. Gash, C. D. Smith, G. A. Gerhardt, R. Kryscio, H. Chebrolu, et al. 2007. "Unilateral intraputamenal glial cell line-derived neurotrophic factor in patients with Parkinson disease: Response to 1 year of treatment and 1 year of withdrawal." *Neurosurgery* 106:614–620.

Sloviter, R. S. 1989. "Calcium-binding protein (calbindin-D28k) and parvalbumin immunocytochemistry: Localization in the rat hippocampus with specific reference to the selective vulnerability of hippocampal neurons to seizure activity." *Journal of Comparative Neurology* 280:183–196.

Smrt, S. D., and X. Zhao. 2010. "Epigenetic regulation of neuronal dendrite and dendritic spine development." *Frontiers in Biology* 5:304–323.

Smith, D. H., K. Okiyama, M. J. Thomas, and T. K. McIntosh. 1993. "Effects of the excitatory amino acid receptor antagonists kynurenate and indole-2-carboxylic acid on behavioral and neurochemical outcome following experimental brain injury." *Journal of Neuroscience* 13:5383–5392.

Smith, O. 1998. "Nobel prize for NO research." *Nature Medicine* 4:1215.

Smith-Swintosky, V. L., L. C. Pettigrew, R. M. Sapolsky, C. Phares, S. D. Craddock, S. M. Brooke, et al. 1996. "Metyrapone, an inhibitor of glucocorticoid production, reduces brain injury induced by focal and global ischemia and seizures." *Journal of Cerebral Blood Flow and Metabolism* 16:585–598.

Soares, J. C., and R. B. Innis. 1999. "Neurochemical brain imaging investigations of schizophrenia." *Biological Psychiatry* 46:600–615.

Spencer, P. S. 2022. "Parkinsonism and motor neuron disorders: Lessons from Western Pacific ALS/PDC." *Journal of Neurological Science* 433:120021. *y.* 47 (supplement 2):S86–90.

Stanek, L. M., S. P. Sardi, B. Mastis, A. R. Richards, C. M. Treleaven, T. Taksir, et al. 2014. "Silencing mutant huntingtin by adeno-associated virus-mediated RNA interference ameliorates disease manifestations in the YAC128 mouse model of Huntington's disease." *Human Gene Therapy* 25:461–474.

Stayte, S., K. J. Laloli, P. Rentsch, A. Lowth, K. M. Li, R. Pickford, et al. 2020. "The kainate receptor antagonist UBP310 but not single deletion of GluK1, GluK2, or GluK3 subunits, inhibits MPTP-induced degeneration in the mouse midbrain." *Experimental Neurology* 323. doi: 10.1016/j.expneurol.2019.113062.

Stein-Behrens, B., M. P. Mattson, I. Chang, M. Yeh, and R. J. Saplosky. 1994. "Stress exacerbates neuron loss and cytoskeletal pathology in the hippocampus." *Journal of Neuroscience* 14:5373–5380.

Stellwagen, D., and R. C. Malenka. 2006. "Synaptic scaling mediated by glial TNF-alpha." *Nature* 440:1054–1059.

Stephan, A. H., B. A. Barres, and B. Stevens. 2012. "The complement system: An unexpected role in synaptic pruning during development and disease." *Annual Review of Neuroscience* 35:369–389.

Stern, Y., A. MacKay-Brandt, S. Lee, P. McKinley, K. McIntyre, Q. Razlighi, et al. 2019. "Effect of aerobic exercise on cognition in younger adults: A randomized clinical trial." *Neurology* 92:e905–e916.

Stranahan, A. M., T. V. Arumugam, R. G. Cutler, K. Lee, J. M. Egan, and M. P. Mattson. 2008. "Diabetes impairs hippocampal function through glucocorticoid-mediated effects on new and mature neurons." *Nature Neuroscience* 11:309–317.

Stranahan, A. M., T. V. Arumugam, K. Lee, and M. P. Mattson. 2010. "Mineralocorticoid receptor activation restores medial perforant path LTP in diabetic rats." *Synapse* 64:528–532.

Stranahan, A. M., K. Lee, B. Martin, S. Maudsley, E. Golden, R. G. Cutler, et al. 2009. "Voluntary exercise and caloric restriction enhance hippocampal dendritic spine density and BDNF levels in diabetic mice." *Hippocampus* 19:951–961.

Svensson, E., E. Horvath-Puho, R. W. Thomsen, J. C. Djurhuus, L. Pedersen, P. Borghammer, et al. 2015. "Vagotomy and subsequent risk of Parkinson's disease." *Annals of Neurology* 78:522–529.

Szepetowski, P. 2018. "Genetics of human epilepsies: Continuing progress." *Presse Medicine* 47:218–226.

Tai, X. Y., M. Koepp, J. S. Duncan, N. Fox, P. Thompson, S. Baxendale, et al. 2016. "Hyperphosphorylated Tau in patients with refractory epilepsy correlates with cognitive decline: A study of temporal lobe resections." *Brain* 139:2441–2455.

Takagaki, G. 1996. "The dawn of excitatory amino acid research in Japan: The pioneering work by Professor Takashi Hayashi." *Neurochemistry International* 29:225–229.

Tan, K. R., U. Rudolph, and C. Luscher. 2011. "Hooked on benzodiazepines: GABAA receptor subtypes and addiction." *Trends in Neurosciences* 34:188–197.

Todd, E. C. D. 1993. "Domoic acid and amnesic shellfish poisoning—a review." *Journal of Food Protection* 56:69–83.

Tomlinson, L., C. V. Leiton, and H. Colognato. 2016. "Behavioral experiences as drivers of oligodendrocyte lineage dynamics and myelin plasticity." *Neuropharmacology* 110:548–562.

Tortarolo, M., G. Grignaschi, N. Calvaresi, E. Zennaro, G. Spaltro, M. Colovic, et al. 2006. "Glutamate AMPA receptors change in motor neurons of SOD1G93A transgenic mice and their inhibition by a noncompetitive antagonist ameliorates the progression of amytrophic lateral sclerosis-like disease." *Journal of Neuroscience Research* 83:134–146.

Toyota, M., D. Spencer, S. Sawai-Toyota, W. Jiaqi, T. Zhang, A. J. Koo, et al. 2018. "Glutamate triggers long-distance, calcium-based plant defense signaling." *Science* 361:1112–1115.

Trudler, D., S. Sanz-Blasco, Y. S. Eisele, S. Ghatak, K. Bodhinathan, M. W. Akhtar, et al. 2021. "Alpha-synuclein oligomers induce glutamate release from astrocytes and excessive extrasynaptic NMDAR activity in neurons, thus contributing to synapse loss." *Journal of Neuroscience* 41:2264–2273.

Tucker, D., Y. Liu, and Q. Zhang. 2018. "From mitochondrial function to neuroprotection—an emerging role for methylene blue." *Molecular Neurobiology* 55:5137–5153.

Uno, Y., and J. T. Coyle. 2019. "Glutamate hypothesis in schizophrenia." *Psychiatry and Clinical Neuroscience* 73:204–215.

Van der Vlag, M., R. Havekes, and P. R. A. Heckman. 2020. "The contribution of Parkin, PINK1 and DJ-1 genes to selective neuronal degeneration in Parkinson's disease." *Journal of Neuroscience* 52:3256–3268.

Vanevski, F., and B. Xu. 2013. "Molecular and neural bases underlying roles of BDNF in the control of body weight." *Frontiers in Neuroscience* 7. doi: 10.3389/fnins.2013.00037.

Verhaeghe, R., V. Gao, S. Morley-Fletcher, H. Bouwalerh, G. Van Camp, F. Cisani, et al. 2021. "Maternal stress programs a demasculinization of glutamatergic transmission in stress-related brain regions of aged rats." *Geroscience* 13:1–23.

Villumsen, M., S. Aznar, B. Pakkenberg, T. Jess, and T. Brudek. 2019. "Inflammatory bowel disease increases the risk of Parkinson's disease: A Danish nationwide cohort study 1977–2014." *Gut* 68:18–24.

Vivar, C., M. C. Potter, J. Choi, J. Y. Lee, T. P. Stringer, E. M. Callaway, et al. 2012. "Monosynaptic inputs to new neurons in the dentate gyrus." *Nature Communications* 3:1107.

Volkow, N. D., and M. Morales. 2015. "The brain on drugs: From reward to addiction." *Cell* 162:712–725.

Von Bartheld, C. S., J. Bahney, and S. Herculano-Houzel. 2016. "The search for true numbers of neurons and glial cells in the human brain: A review of 150 years of cell counting." *Journal of Comparative Neurology* 524:3865–3895.

Voss, M. W., C. Soto, S. Yoo, M. Sodoma, C. Vivar, and H. van Praag. 2019. "Exercise and hippocampal memory systems." *Trends in Cognitive Sciences* 23:318–333.

Wan, R., L. A. Weigand, R. Bateman, K. Griffioen, D. Mendelowitz, and M. P. Mattson. 2014. "Evidence that BDNF regulates heart rate by a mechanism involving increased brainstem parasympathetic neuron excitability." *Journal of Neurochemistry* 129:573–580.

Watkins, J. C., and D. E. Jane. 2006. "The glutamate story." *British Journal of Pharmacology* 147:S100–S108.

Weidemann, A., G. Konig, D. Bunke, P. Fischer, J. M. Salbaum, C. L. Masters, et al. 1989. "Identification, biogenesis, and localization of precursors of Alzheimer's disease A4 amyloid protein." *Cell* 57:115–126.

Weizmann, L., L. Dayan, S. Brill, H. Nahman-Averbuch, T. Hendler, G. Jacob, et al. 2018. "Cannabis analgesia in chronic neuropathic pain is associated with altered brain connectivity." *Neurology* 91:e1285–e1294.

Whone, A., M. Luz, M. Boca, M. Woolley, L. Mooney, S. Dharia, et al. 2019. "Randomized trial of intermittent intraputamenal glial cell line-derived neurotrophic factor in Parkinson's disease." *Brain* 142:512–525.

Williams, T. I., B. C. Lynn, W. R. Markesbery, and M. A. Lovell. 2006. "Increased levels of 4-hydroxynonenal and acrolein, neurotoxic markers of lipid peroxidation, in the brain in mild cognitive impairment and early Alzheimer's disease." *Neurobiology of Aging* 27:1094–1099.

Wilson, R. I., and R. A. Nicoll. 2002. "Endocannabinoid signaling in the brain." *Science* 296:678–682.

Winden, K. D., D. Ebrahimi-Fakhari, and M. Shahin. 2018. "Abnormal mTOR activation in autism." *Annual Review of Neuroscience* 41:1–23.

Witkin, J. M., J. Kranzler, K. Kaniecki, P. Popik, J. L. Smith, K. Hashimoto, et al. 2020. "R-(-)-ketamine modifies behavioral effects of morphine predicting efficacy as a novel therapy for opioid use disorder." *Journal of Pharmacology Biochemistry and Behavior* 194. doi: 10.1016/j.pbb.2020.172927.

Wofsey, A. R., M. J. Kuhar, and S. H. Snyder. 1971. "A unique synaptosomal fraction, which accumulates glutamic and aspartic acids, in brain tissue." *Proceedings of the National Academy of Sciences, USA* 68:1102–1106.

Wong, F. K., and O. Marin. 2019. "Developmental cell death in the cerebral cortex." *Annual Review of Cell and Developmental Biology* 35:523–542.

Worrell, S. D., and T. J. Gould. 2021. "Therapeutic potential of ketamine for alcohol use disorder." *Neuroscience and Biobehavioral Reviews* 126:573–589.

Wu, B., M. Jiang, Q. Peng, G. Li, Z. Hou, G. L. Milne, et al. 2017. "2,4 DNP improves motor function, preserves medium spiny neuronal identity, and reduces oxidative stress in a mouse model of Huntington's disease." *Experimental Neurology* 293:83–90.

Wu, J. W., S. A. Hussaini, I. M. Bastille, G. A. Rodriguez, A. Mrejeru, K. Rilett, et al. 2016. "Neuronal activity enhances Tau propagation and Tau pathology in vivo." *Nature Neuroscience* 19:1085–1092.

Wu, Y., and C. Janetopoulos. 2013. "Systematic analysis of gamma-aminobutyric acid (GABA) metabolism and function in the social amoeba *Dictyostelium discoideum*." *Journal of Biological Chemistry* 288:15280–15290.

Xiong, M., O. D. Jones, K. Peppercorn, S. M. Ohline, W. P. Tate, and W. C. Abraham. 2017. "Secreted amyloid precursor protein-alpha can restore novel object location memory and hippocampal LTP in aged rats." *Neurobiology of Learning and Memory* 138:n291–n299.

Yang, J. L., T. Tadokoro, G. Keijzers, M. P. Mattson, and V. A. Bohr. 2010. "Neurons efficiently repair glutamate-induced oxidative DNA damage by a process involving CREB-mediated up-regulation of apurinic endonuclease 1." *Journal of Biological Chemistry* 285:28191–28199.

Yang, X., F. Tian, H. Zhang, J. Zeng, T. Chen, S. Wang, et al. 2016. "Cortical and subcortical gray matter shrinkage in alcohol-use disorders: A voxel-based meta-analysis." *Neuroscience Biobehavioral Reviews* 66:92–103.

Yau, P. L., M. G. Castro, A. Tagani, W. H. Tsui, and A. Convit. 2012. "Obesity and metabolic syndrome and functional and structural brain impairments in adolescence." *Pediatrics* 130:e856–e864.

Yu, Z. F., and M. P. Mattson. 1999. "Dietary restriction and 2-deoxyglucose administration reduce focal ischemic brain damage and improve behavioral outcome: Evidence for a preconditioning mechanism." *Journal of Neuroscience Research* 57:830–839.

Yuan, H., C. M. Low, O. A. Moody, A. Jenkins, and S. F. Traynelis. 2015. "Ionotropic GABA and glutamate receptor mutations and human neurologic diseases." *Molecular Pharmacology* 88:203–217.

Zarate, C. A., Jr., N. E. Brutsche, L. Ibrahim, J. Franco-Chaves, N. Diazgranados, A. Cravchik, et al. 2012. "Replication of ketamine's antidepressant efficacy in bipolar depression: A randomized controlled add-on trial." *Biological Psychiatry* 71:939–946.

Zarate, C. A., Jr., J. B. Singh, P. J. Carlson, N. E. Brutsche, R. Ameli, D. A. Luckenbaugh, et al. 2006. "A randomized trial of an N-methyl-D-aspartate antagonist in treatment-resistant major depression." *Archives of General Psychiatry* 63:856–864.

Zayed, A., and G. E. Robinson. 2012. "Understanding the relationship between brain gene expression and social behavior: Lessons from the honey bee." *Annual Review of Genetics* 46:591–615.

Zeithamova, D., M. L. Schlichting, and A. R. Preson. 2012. "The hippocampus and inferential reasoning: Building memories to navigate future decisions." *Frontiers in Human Neuroscience* 6. doi: 10.3389/fnhum.2012.00070.

Zhang, S., E. Eitan, T. Y. Wu, and M. P. Mattson. 2018. "Intercellular transfer of pathogenic alpha-synuclein by extracellular vesicles is induced by the lipid peroxidation product 4-hydroxynonenal." *Neurobiology of Aging* 61:52–65.

Zhao, M., D. Li, K. Shimazu, Y. X. Zhou, B. Lu, and C. X. Deng. 2007. "Fibroblast growth factor receptor-1 is required for long-term potentiation, memory consolidation, and neurogenesis." *Biological Psychiatry* 62:381–390.

Zuccato, C., M. Tartari, A. Crotti, D. Goffredo, M. Valenza, L. Conti, et al. 2003. "Huntingtin interacts with REST/NRSF to modulate the transcription of NRSE-controlled neuronal genes." *Nature Genetics* 35:76–83.

Further Reading

CHAPTER 2

Feehily, C., and K. A. G. Karatzas. 2013. "Role of glutamate metabolism in bacterial responses towards acid and other stresses." *Journal of Applied Microbiology* 114:11–24.

Luedtke, S., V. O'Connor, L. Holden-Dye, and R. J. Walker. 2010. "The regulation of feeding and metabolism in response to food deprivation in *Caenorhabditis elegans*." *Journal of Invertebrate Neuroscience* 10:63–76.

Mattson, M. P. 2019. "An evolutionary perspective on why food overconsumption impairs cognition." *Trends in Cognitive Science* 23:200–212.

Mattson, M. P. 2022. *The Intermittent Fasting Revolution: The Science of Optimizing Health and Enhancing Performance*. Cambridge, MA: MIT Press.

Moroz, L. L., M. A. Nikitin, P. G. Policar, A. B. Kohn, and D. Y. Romanova. 2021. "Evolution of glutamatergic signaling and synapses." *Neuropharmacology* 199. doi: 10.1016/j.neuropharm.2021.108740.

O'Rourke, T., and C. Boeckx. 2020. "Glutamate receptors in domestication and modern human evolution." *Neuroscience Biobehavioral Reviews* 108:341–357.

Ramos-Vicente, D., S. G. Grant, and A. A. Bayes. 2021. "Metazoan evolution and diversity of glutamate receptors and their auxiliary subunits." *Neuropharmacology* 195. doi: 10.1016/j.neuropharm.2021.108640.

Sotelo, C., and I. Dusart. 2009. "Intrinsic versus extrinsic determinants during the development of Purkinje cell dendrites." *Neuroscience* 162:589–600.

Streidter, G. F. 2005. *Principles of Brain Evolution*. Sutherland, MA: Sinauer Associates.

Volkow, N. D., R. A. Wise, and R. Baler. 2017. "The dopamine motive system: Implications for drug and food addiction." *Nature Reviews Neuroscience* 18:741–752.

Watkins, J. C., and D. E. Jane. 2006. "The glutamate story." *British Journal of Pharmacology* 147:S100–S108.

Young, V. R., and A. M. Ajami. 2000. "Glutamate: An amino acid of particular distinction." *Journal of Nutrition* 130:892S–900S.

CHAPTER 3

Cline, H., and K. Haas. 2008. "The regulation of dendritic arbor development and plasticity by glutamatergic synaptic input: A review of the synaptotrophic hypothesis." *Journal of Physiology* 586:1509–1517.

Gibb, R., and B. Kolb. 2014. *The Neurobiology of Brain and Behavioral Development.* New York: Academic Press.

Habermacher, C., M. C. Angulo, and N. Benamer. 2019. "Glutamate versus GABA in neuron-oligodendroglia communication." *Glia* 67:2092–2106.

Mattson, M. P. 1988. "Neurotransmitters in the regulation of neuronal cytoarchitecture." *Brain Research* 472:179–212.

Reemst, K., S. C. Noctor, P. J. Lucassen, and E. M. Hol. 2016. "The indispensable roles of microglia and astrocytes during brain development." *Frontiers in Human Neuroscience* 10. doi: 10.3389/fnhum.2016.00566.

Sans, D., T. Reh, W. Harris, and M. Landgraf. 2019. *Development of the Nervous System.* 4th ed. New York: Academic Press.

CHAPTER 4

Andersen, P., R. Morris, D. Amaral, T. Bliss, and J. O'Keefe. 2007. *The Hippocampus Book.* Oxford: Oxford University Press.

Andersson, K. E. 2018. "PDE5 inhibitors—pharmacology and clinical applications 20 years after sildenafil discovery." *British Journal of Pharmacology* 175:2554–2565.

Bliss, T. V., and G. L. Collingridge. 1993. "A synaptic model of memory: Long-term potentiation in the hippocampus." *Nature* 361:31–39.

Chaaya, N., A. R. Battle, and L. R. Johnson. 2018. "An update on contextual fear memory mechanisms: Transition between amygdala and hippocampus." *Neuroscience and Biobehavioral Research* 92:43–54.

Coglin, L. L. 2013. "Mechanisms and functions of theta rhythms." *Annual Review of Neuroscience* 36:295–312.

Epstein, R. A., E. Z. Patai, J. B. Julian, and H. J. Spiers. 2017. "The cognitive map in humans: Spatial navigation and beyond." *Nature Neuroscience* 20:1504–1513.

Leal, G., D. Comprido, and C. B. Duarte. 2014. "BDNF-induced local protein synthesis and synaptic plasticity." *Neuropharmacology* 76:639–656.

Lieberman, D. A. 2012. *Learning and Memory.* Cambridge: Cambridge University Press.

Theves, S., G. Fernandez, and C. F. Doeller. 2019. "The hippocampus encodes distances in multidimensional feature space." *Current Biology* 29:1226–1231.

CHAPTER 5

Camandola, C., and M. P. Mattson. 2017. "Brain metabolism in health, aging, and neurodegeneration." *EMBO Journal* 36:1474–1492.

Golgi, C. 1886. *Sulla fina anatomia degli organi centrali del sistema nervoso*. Milan: Hoepli.

Iadecola, C., and M. Nedergaard. 2007. "Glial regulation of the cerebral microvasculature." *Nature Neuroscience* 10:1369–1376.

Kerr, J. S., B. A. Adriaanse, N. H. Greig, M. P. Mattson, M. Z. Cader, V. A. Bohr, et al. 2017. "Mitophagy and Alzheimer's disease: Cellular and molecular mechanisms." *Trends in Neurosciences* 40:151–166.

Liu, Y., A. Cheng, Y. J. Li, Y. Yang, Y. Kishimoto, S. Wang, et al. 2019. "SIRT3 mediates hippocampal synaptic adaptations to intermittent fasting and ameliorates deficits in APP mutant mice." *Nature Communications* 10. doi: 10.1038/s41467-019-09897-1.

Magistretti, P. J., and I. Allaman. 2015. "A cellular perspective on brain energy metabolism and functional imaging." *Neuron* 86:883–901.

Mattson, M. P. 2022. *The Intermittent Fasting Revolution: The Science of Optimizing Health and Enhancing Performance*. Cambridge, MA: MIT Press.

Rothman, S. M., K. J. Griffioen, R. Wan, and M. P. Mattson. 2012. "Brain-derived neurotrophic factor as a regulator of systemic and brain energy metabolism and cardiovascular health." *Annals of the New York Academy of Sciences* 1264:49–63.

CHAPTER 6

Blaylock, R. L. 2011. *Excitotoxins: The Taste That Kills*. Santa Fe, NM: Health Press.

Costa, L. G., G. Giordano, and E. M. Faustman. 2010. "Domoic acid as a developmental neurotoxin." *Neurotoxicology* 31:409–423.

Coyle, J. T. 1987. "Kainic acid: Insights into excitatory mechanisms causing selective neuronal degeneration." *Ciba Foundation Symposium* 126:186–203.

Duty, S., and P. Jenner. 2011. "Animal models of Parkinson's disease: A source of novel treatments and clues to the cause of the disease." *British Journal of Pharmacology* 164:1357–1391.

Glazner, G. W., S. L. Chan, C. Lu, and M. P. Mattson. 2000. "Caspase-mediated degradation of AMPA receptor subunits: A mechanism for preventing excitotoxic necrosis and ensuring apoptosis." *Journal of Neuroscience* 20:3641–3649.

Mattson, M. P. 2008. "Glutamate and neurotrophic factors in neuronal plasticity and disease." *Annals of the New York Academy of Science* 1144:97–112.

Mattson, M. P., and W. Duan. 1999. "'Apoptotic' biochemical cascades in synaptic compartments: Roles in adaptive plasticity and neurodegenerative disorders." *Journal of Neuroscience Research* 58:152–166.

Sloviter, R. S. 1989. "Calcium-binding protein (calbindin-D28k) and parvalbumin immunocyto-chemistry: Localization in the rat hippocampus with specific reference to the selective vulnerability of hippocampal neurons to seizure activity." *Journal of Comparative Neurology* 280:183–196.

CHAPTER 7

Casault, A. I., A. S. Sultan, M. Banoei, P. Couillar, A. Kramer, and B. W. Winston. 2019. "Cytokine responses in severe traumatic brain injury: Where there is smoke, is there fire?" *Neurocritical Care* 30:22–32.

Collins, R. C., B. H. Dobkin, and D. W. Choi. 1989. "Selective vulnerability of the brain: New insights into the pathophysiology of stroke." *Annals of Internal Medicine* 110:992–1000.

Krishnamurthy, K., and D. T. Laskowitz. 2016. "Cellular and molecular mechanisms of secondary neuronal injury following traumatic brain injury." In *Translational Research in Traumatic Brain Injury*, edited by D. Laskowitz and G. Grant, 97–126. Boca Raton, FL: CRC Press/Taylor Francis Group.

Kruman, I. I., W. A. Pedersen, J. E. Springer, and M. P. Mattson. 1999. "ALS-linked Cu/Zn-SOD mutation increases vulnerability of motor neurons to excitotoxicity by a mechanism involving increased oxidative stress and perturbed calcium homeostasis." *Experimental Neurology* 160: 28–39.

Liu, Y., L. J. Zhou, J. Wang, D. Li, W. J. Ren, J. Peng, et al. 2017. "TNF-alpha differentially regulates synaptic plasticity in the hippocampus and spinal cord by microglia-dependent mechanisms after peripheral nerve injury." *Journal of Neuroscience* 37:871–881.

Lomazow, S. 2011. "The epilepsy of Franklin Delano Roosevelt." *Neurology* 76:668–669.

Ludhiadch, A., R. Sharma, A. Muriki, and A. Munshi. 2022. "Role of calcium homeostasis in ischemic stroke: A review." *CNS & Neurological Disorders—Drug Targets* 21:52–61.

Park, E., A. A. Velumian, and M. G. Fehlings. 2004. "The role of excitotoxicity in secondary mechanisms of spinal cord injury: A review with an emphasis on the implications for white matter degeneration." *Journal of Neurotrauma* 21:754–774.

Prins, M., T. Greco, D. Alexander, and C. C. Giza. 2013. "The pathophysiology of traumatic brain injury at a glance." *Disease Models and Mechanisms* 6:1307–1315.

Wu, Q. J., and M. Tymianski. 2018. "Targeting NMDA receptors in stroke: New hope in neuroprotection." *Molecular Brain* 11. doi: 10.1186/s13041-018-0357-8.

CHAPTER 8

Blasco, H., S. Mavel, P. Corcia, and P. H. Gordon. 2014. "The glutamate hypothesis in ALS: Pathophysiology and drug development." *Current Medicinal Chemistry* 21:3551–3575.

Dash, D., and T. A Mestre. 2020. "Therapeutic update on Huntington's disease: Symptomatic treatments and emerging disease-modifying therapies." *Neurotherapeutics* 17:1645–1659.

Doyle, M. E., and J. M. Egan. 2001. "Glucagon-like peptide-1." *Recent Progress in Hormone Research* 56:377–399.

Evans, J. R., and R. A. Barker. 2008. "Neurotrophic factors as a therapeutic target for Parkinson's disease." *Expert Opinion in Therapeutic Targets* 12:437–447.

Fan, M. M., and L. A. Raymond. 2007. "N-methyl-D-aspartate (NMDA) receptor function and excitotoxicity in Huntington's disease." *Progress in Neurobiology* 81:272–293.

Jones, L., and A. Hughes. 2011. "Pathogenic mechanisms in Huntington's disease." *International Review of Neurobiology* 98:373–418.

Lee, H. S., E. Lobbestael, S. Vermeire, J. Sabino, and I. Cleynen. 2021. "Inflammatory bowel disease and Parkinson's disease: Common pathophysiological links." *Gut* 70:408–417.

Martinen, M., M. Takalo, T. Natunen, R. Wittrahm, S. Gabouj, S. Kemppainen, et al. 2018. "Molecular mechanisms of synaptotoxicity and neuroinflammation in Alzheimer's disease." *Frontiers in Neuroscience* 12. doi: 10.3389/fnins.2018.00963.

Olanow, C. W. 2007. "The pathogenesis of cell death in Parkinson's disease—2007." *Movement Disorders* 17:S335–S342.

Ong, W. Y., K. Tanaka, G. S. Dawe, L. M. Ittner, and A. A. Farooqui. 2013. "Slow excitotoxicity in Alzheimer's disease." *Journal of Alzheimer's Disease* 35:643–648.

Raskin, J., J. Cummings, J. Hardy, K. Schuh, and R. A. Dean. 2015. "Neurobiology of Alzheimer's disease: Integrated molecular, physiological, anatomical, biomarker, and cognitive dimensions." *Current Alzheimer's Research* 12:712–722.

Ryan, B. J., S. Hock, E. A. Fon, and R. Wade-Martins. 2015. "Mitochondrial dysfunction and mitophagy in Parkinson's: From familial to sporadic disease." *Trends in Biochemical Sciences* 40:200–210.

Savitt, J. M., V. L. Dawson, and T. M. Dawson. 2006. "Diagnosis and treatment of Parkinson [*sic*] disease: Molecules to medicine." *Journal of Clinical Investigation* 116:1744–1754.

Sepers, M. S., and L. A. Raymond. 2014. "Mechanisms of synaptic dysfunction and excitotoxicity in Huntington's disease." *Drug Discovery Today* 19:990–996.

Van Den Bosch, L., P. Van Damme, E. Bogaert, and W. Robberecht. 2006. "The role of excitotoxicity in the pathogenesis of amyotrophic lateral sclerosis." *Biochimica et Biophysica Acta* 1762:1068–1082.

Zuccato, C., and E. Cattaneo. 2007. "Role of brain derived neurotrophic factor in Huntington's disease." *Progress in Neurobiology* 81:294–330.

CHAPTER 9

Averill, L. A., P. Purohit, C. L. Averill, M. A. Boesl, J. H. Krystal, and C G. Abdallah. 2017. "Glutamate dysregulation and glutamatergic therapeutics for PTSD: Evidence from human studies." *Neuroscience Letters* 649:147–155.

Charney, D. S., E. J. Nestler, P. Skylar, and J. D. Buxbaum. 2018. *Neurobiology of Mental Illness.* 5th ed. New York: Oxford University Press.

Duman, R. S., and L. M. Monteggia. 2006. "A neurotrophic model for stress-related mood disorders." *Biological Psychiatry* 59:1116–1127.

Grandin, T. 2005. *Animals in Translation: Using the Mysteries of Autism to Decode Animal Behavior.* New York: Harcourt.

Kamiya, K., and O. Abe. 2020. "Imaging of posttraumatic stress disorder." *Neuroimaging Clinics of North America* 30:115–123.

Kondziella, D., S. Alvestad, A. Vaalar, and U. J. Sonnewald. 2007. "Which clinical and experimental data link temporal lobe epilepsy with depression?" *Journal of Neurochemistry* 103: 136–152.

Koo, J. W., D. Chaudhury, M. H. Han, and E. J. Nestler. 2019. "Role of mesolimbic brain-derived neurotrophic factor in depression." *Biological Psychiatry* 86:738–748.

Lasarge, C. L., and S. C. Danzer. 2014. "Mechanisms regulating neuronal excitability and seizure development following mTOR pathway hyperactivation." *Frontiers in Molecular Neuroscience* 14. doi: 10.3389/fnmol.2014.00018.

Nasir, M., D. Trujillo, J. Levine, J. B. Dwyer, Z. W. Rupp, and M. H. Bloch. 2020. "Glutamate systems in DSM-5 anxiety disorders: Their role and a review of glutamate and GABA psychopharmacology." *Frontiers in Psychiatry* 11. doi: 10.3389/fpsyt.2020.548505.

Ploski, J. E., and V. A. Vaidya. 2021. "The neurocircuitry of posttraumatic stress disorder and major depression: Insights into overlapping and distinct circuit dysfunction—a tribute to Ron Duman." *Biological Psychiatry* 90:109–117.

Stranahan, A. M., T. V. Arumugam, R. G. Cutler, K. Lee, J. M. Egan, and M. P. Mattson. 2008. "Diabetes impairs hippocampal function through glucocorticoid-mediated effects on new and mature neurons." *Nature Neuroscience* 11:309–317.

Zanos, P., and T. D. Gould. 2018. "Mechanisms of ketamine action as an antidepressant." *Molecular Psychiatry* 23:801–811.

CHAPTER 10

Barker, S. A. 2018. "N, N-dimethyltryptamine (DMT), an endogenous hallucinogen: Past, present, and future research to determine its role and function." *Frontiers in Neuroscience* 12. doi: 10.3389/fnins.2018.00536.

Carhart-Harris, R. L., D. Erritzoe, T. Williams, J. M. Stone, L. J. Reed, A. Colasanti, et al. 2012. "Neural correlates of the psychedelic state as determined by fMRI studies with psilocybin." *Proceedings of the National Academy of Sciences USA* 109:2138–2143.

De Gregorio, D., A. Aguilar-Valles, K. H. Preller, B. D. Heifets, M. Hibicke, J. Mitchell, et al. 2021. "Hallucinogens in mental health: Preclinical and clinical studies on LSD, psilocybin, MDMA, and ketamine." *Journal of Neuroscience* 41:891–900.

Duman, R. S., G. Sanacora, and J. H. Krystal. 2019. "Altered connectivity in depression: GABA and glutamate neurotransmitter deficits and reversal by novel treatments." *Neuron* 102:75–90.

Engin, E., R. R. Benham, and U. Rudolph. 2018. "An emerging circuit pharmacology of GABAA receptors." *Trends in Pharmacological Sciences* 39:710–732.

Heckers, S., and C. Konradi. 2015. "GABAergic mechanisms of hippocampal hyperactivity in schizophrenia." *Schizophrenia Research* 167:4–11.

Jones, J. L., C. F. Mateus, R. J. Malcolm, K. T. Brady, and S. E. Back. 2018. "Efficacy of ketamine in the treatment of substance use disorders: A systematic review." *Frontiers in Psychiatry* 9. doi: 10.3389/fpsyt.2018.00277.

Koob, G. F., and N. D. Volkow. 2016. "Neurobiology of addiction: A neurocircuitry analysis." *Lancet Psychiatry* 3:760–773.

Leary, T. 1980. *The Politics of Ecstasy*. Berkeley, CA: Ronin.

Pollan, M. 2018. *How to Change Your Mind*. New York: Penguin.

Russo, S. J., D. M. Dietz, D. Dumitriu, J. H. Morrison, R. C. Malenka, and E. J. Nestler. 2010. "The addicted synapse: Mechanisms of synaptic and structural plasticity in nucleus accumbens." *Trends in Neurosciences* 33:267–276.

Vollenwider, F. X., and M. Kometer. 2010. "The neurobiology of psychedelic drugs: Implications for the treatment of mood disorders." *Nature Reviews Neuroscience* 11:642–651.

CHAPTER 11

Cunnane, S. C., E. Trushina, C. Morland, A. Prigione, G. Casadesus, Z. B. Andrews, et al. 2020. "Brain energy rescue: An emerging therapeutic concept for neurodegenerative disorders of ageing." *Nature Reviews Drug Discovery* 19:609–633.

Kellar, D., and S. Craft. 2020. "Brain insulin resistance in Alzheimer's disease and related disorders: Mechanisms and therapeutic approaches." *Lancet Neurology* 19:758–766.

Marosi, K., and M. P. Mattson. 2014. "BDNF mediates adaptive brain and body responses to energetic challenges." *Trends in Endocrinology and Metabolism* 25:89–98.

Mattson, M. P. 2015. "Lifelong brain health is a lifelong challenge: From evolutionary principles to empirical evidence." *Ageing Research Reviews* 20:37–45.

Mattson, M. P. 2022. *The Intermittent Fasting Revolution: The Science of Optimizing Health and Enhancing Performance*. Cambridge, MA: MIT Press.

Miller, A. A., and S. J. Spencer. 2014. "Obesity and neuroinflammation: A pathway to cognitive impairment." *Brain Behavior and Immunity* 42:10–21.

Phillips, C., M. A. Baktir, M. Srivatsan, and A. Salehi. 2014. "Neuroprotective effects of physical activity on the brain: A closer look at trophic factor signaling." *Frontiers in Cellular Neuroscience* 8. doi: 10.3389/fncel.2014.00170.

Stranahan, A. M., and M. P. Mattson. 2012. "Recruiting adaptive cellular stress responses for successful brain ageing." *Nature Reviews Neuroscience* 13:209–216.

Index